W0227669

Orthogonal Polynomials and Painlevé Equations

There are a number of intriguing connections between Painlevé equations and orthogonal polynomials, and this book is one of the first to provide an introduction to these. Researchers in integrable systems and nonlinear equations will find the many explicit examples where Painlevé equations appear in mathematical analysis very useful. Those interested in the asymptotic behavior of orthogonal polynomials will also find the description of Painlevé transcendents and their use for local analysis near certain critical points helpful to their work. Rational solutions and special function solutions of Painlevé equations are worked out in detail, with a survey of recent results and an outline of their close relationship with orthogonal polynomials. Exercises throughout the book help the reader to get to grips with the material.

The author is a leading authority on orthogonal polynomials, giving this work a unique perspective on Painlevé equations.

AUSTRALIAN MATHEMATICAL SOCIETY LECTURE SERIES

Editor-in-chief: Professor J. Ramagge, School of Mathematics and Statistics, University of Sydney, NSW 2006, Australia

Editors:
Professor G. Froyland, School of Mathematics and Statistics, University of New South Wales, NSW 2052, Australia

Professor M. Murray, School of Mathematical Sciences, University of Adelaide, SA 5005, Australia

Professor C. Praeger, School of Mathematics and Statistics, University of Western Australia, Crawley, WA 6009, Australia

Australian Mathematical Society Lecture Series: 27

Orthogonal Polynomials and Painlevé Equations

WALTER VAN ASSCHE

Katholieke Universiteit Leuven, Belgium

CAMBRIDGE
UNIVERSITY PRESS

CAMBRIDGE
UNIVERSITY PRESS

University Printing House, Cambridge CB2 8BS, United Kingdom

One Liberty Plaza, 20th Floor, New York, NY 10006, USA

477 Williamstown Road, Port Melbourne, VIC 3207, Australia

314–321, 3rd Floor, Plot 3, Splendor Forum, Jasola District Centre,
New Delhi – 110025, India

79 Anson Road, #06–04/06, Singapore 079906

Cambridge University Press is part of the University of Cambridge.

It furthers the University's mission by disseminating knowledge in the pursuit of
education, learning, and research at the highest international levels of excellence.

www.cambridge.org
Information on this title: www.cambridge.org/9781108441940
DOI: 10.1017/9781108644860

First published 2018

Printed in the United Kingdom by Clays, St Ives plc

A catalogue record for this publication is available from the British Library.

ISBN 978-1-108-44194-0 Paperback

Contents

Preface

These notes are intended to explain the relationship between orthogonal polynomials and Painlevé equations. They are not intended to give a systematic theory of Painlevé equations, their transformations and classification. This can be found elsewhere; we recommend in particular the classical book by Ince [88, §14.4], the more recent books [46, 68, 82, 125] and the review papers [80] and [96] on discrete and continuous Painlevé equations. Researchers in orthogonal polynomials will find the notes useful to see how semi-classical orthogonal polynomials often lead to discrete and continuous Painlevé equations. Usually only special solutions of these Painlevé equations in terms of classical special functions will be relevant. Furthermore, some integrable systems, such as the Toda lattice and related differential-difference systems, also appear in a very natural way in the theory of orthogonal polynomials. Those interested in the asymptotic behavior of orthogonal polynomials may appreciate seeing that Painlevé transcendents are used for the local analysis near critical points. Researchers in integrable systems, and in particular in Painlevé equations, may find it useful to see that a lot of explicit systems of orthogonal polynomials are described using discrete and continuous Painlevé equations. These applications in orthogonal polynomial theory often give a new viewpoint of Painlevé equations, in particular on the behavior of the special solutions of these equations.

Acknowledgments

I started writing these notes for a master course that I taught at the Universidad Carlos III de Madrid in May 2012. That course essentially consisted of Chapters 1–3. I continued these notes during my sabbatical trip to the University of Sydney in August 2016, where I started Chapters 4 and 6–7. I finally finished the notes at the African Institute for Mathematics in Cameroon in March–April

2017. I would like to thank the people in Leganés (Madrid), Sydney and AIMS Cameroon for their hospitality and for the discussions which I had on the topics of these notes. I am also grateful to the referee for pointing out many papers that are relevant for these lecture notes.

1
Introduction

1.1 Orthogonal polynomials on the real line

Orthonormal polynomials $(p_n)_{n \in \mathbb{N}}$ on the real line are defined by the orthogonality conditions

$$\int_{\mathbb{R}} p_n(x) p_m(x) \, d\mu(x) = \delta_{m,n}, \qquad (1.1)$$

where μ is a positive measure on the real line for which all the moments exist and $p_n(x) = \gamma_n x^n + \cdots$, with positive leading coefficient $\gamma_n > 0$. A family of orthonormal polynomials always satisfies a three-term recurrence relation of the form

$$x p_n(x) = a_{n+1} p_{n+1}(x) + b_n p_n(x) + a_n p_{n-1}(x), \qquad n \geq 0, \qquad (1.2)$$

with $p_{-1} = 0$ and

$$p_0 = \gamma_0 = (\mu(\mathbb{R}))^{-1/2}.$$

Comparing the leading coefficients in the recurrence relation gives

$$a_{n+1} = \frac{\gamma_n}{\gamma_{n+1}} > 0, \qquad (1.3)$$

and computing the Fourier coefficients of $x p_n(x)$ in (1.2) gives

$$a_n = \int_{\mathbb{R}} x p_n(x) p_{n-1}(x) \, d\mu(x), \qquad (1.4)$$

$$b_n = \int_{\mathbb{R}} x p_n^2(x) \, d\mu(x). \qquad (1.5)$$

For the monic orthogonal polynomials $P_n = p_n / \gamma_n$ the recurrence relation is

$$P_{n+1}(x) = (x - b_n) P_n(x) - a_n^2 P_{n-1}(x), \qquad (1.6)$$

where the b_n are as in (1.5) and the a_n^2 are the squares of (1.4).

The converse statement is also true and is known as *the spectral theorem for orthogonal polynomials*[1]: if a family of polynomials satisfies a three-term recurrence relation of the form (1.2), with $a_n > 0$ and $b_n \in \mathbb{R}$ and with initial conditions $p_0 = 1$ and $p_{-1} = 0$, then there exists a probability measure μ on the real line such that these polynomials are orthonormal polynomials satisfying (1.1). This gives rise to two important problems:

Problem 1. Suppose the measure μ is known. What can be said about the recurrence coefficients $(a_n)_{n=1,2,3,\ldots}$ and $(b_n)_{n=0,1,2,\ldots}$? This is known as the *direct problem for orthogonal polynomials*.

Problem 2. Suppose the recurrence coefficients $(a_{n+1}, b_n)_{n=0,1,2,\ldots}$ are known. What can be said about the orthogonality measure μ? This is known as the *inverse problem for orthogonal polynomials*.

A partial solution of problem 1 is that one can express the recurrence coefficients a_n^2 and b_n in terms of the moments of the measure μ. Let

$$m_n = \int_{\mathbb{R}} x^n \, d\mu(x), \qquad n \geq 0,$$

and define the Hankel determinants

$$\Delta_{n+1} = \det \begin{pmatrix} m_0 & m_1 & m_2 & \cdots & m_{n-1} & m_n \\ m_1 & m_2 & m_3 & \cdots & m_n & m_{n+1} \\ m_2 & m_3 & m_4 & \cdots & m_{n+1} & m_{n+2} \\ \vdots & \vdots & \vdots & \cdots & \vdots & \vdots \\ m_n & m_{n+1} & m_{n+2} & \cdots & m_{2n-1} & m_{2n} \end{pmatrix}, \qquad (1.7)$$

then the monic orthogonal polynomial P_n is given by

$$P_n(x) = \frac{1}{\Delta_n} \det \begin{pmatrix} m_0 & m_1 & m_2 & \cdots & m_{n-1} & m_n \\ m_1 & m_2 & m_3 & \cdots & m_n & m_{n+1} \\ m_2 & m_3 & m_4 & \cdots & m_{n+1} & m_{n+2} \\ \vdots & \vdots & \vdots & \cdots & \vdots & \vdots \\ m_{n-1} & m_n & m_{n+1} & \cdots & m_{2n-2} & m_{2n-1} \\ 1 & x & x^2 & \cdots & x^{n-1} & x^n \end{pmatrix}. \qquad (1.8)$$

From this one easily computes

$$\frac{1}{\gamma_n^2} = \int_{\mathbb{R}} P_n^2(x) \, d\mu(x) = \frac{\Delta_{n+1}}{\Delta_n},$$

---

[1] Also known as Favard's theorem, but the result is much older hence the attribution to Favard is not so appropriate.

so that from (1.3) one finds

$$a_n^2 = \frac{\Delta_{n+1}\Delta_{n-1}}{\Delta_n^2}. \qquad (1.9)$$

If we write $P_n(x) = x^n + \delta_n x^{n-1} + \cdots$ and compare the coefficients of x^n in (1.6), then one finds

$$b_n = \delta_n - \delta_{n+1}. \qquad (1.10)$$

The coefficient δ_n can be obtained from (1.8) and is $\delta_n = -\Delta_n^*/\Delta_n$, where Δ_n^* is obtained from Δ_n by replacing the last column $(m_{n-1}, m_n, \ldots, m_{2n-2})^T$ by $(m_n, m_{n+1}, \ldots, m_{2n-1})^T$. One then has from (1.10)

$$b_n = \frac{\Delta_{n+1}^*}{\Delta_{n+1}} - \frac{\Delta_n^*}{\Delta_n}. \qquad (1.11)$$

The formulas (1.9) and (1.11) however do not really show how properties of the measure μ can be transferred to properties of the recurrence coefficients. One needs more tools to solve this direct problem for orthogonal polynomials.

The recurrence coefficients are usually collected in a tridiagonal matrix of the form

$$J = \begin{pmatrix} b_0 & a_1 & 0 & 0 & 0 \\ a_1 & b_1 & a_2 & 0 & 0 \\ 0 & a_2 & b_2 & a_3 & 0 \\ 0 & 0 & a_3 & b_3 & \ddots \\ 0 & 0 & 0 & \ddots & \ddots \end{pmatrix}, \qquad (1.12)$$

which acts as an operator on (a subset of) $\ell^2(\mathbb{N})$ and which is known as the *Jacobi matrix* or Jacobi operator. If J is selfadjoint, then the spectral measure for J is precisely the orthogonality measure μ. Hence problem 1 corresponds to the inverse problem for the Jacobi matrix J and problem 2 corresponds to the direct problem for J.

In the present notes we will study problem 1 for a few special cases. In Chapter 2 we study measures on the real line with an exponential weight function of the form $d\mu(x) = |x|^\rho \exp(-|x|^m)\,dx$, which are known as Freud weights, named after Géza Freud who studied them in the 1970s. It will be shown that the recurrence coefficients $(a_n)_{n \geq 1}$ satisfy a nonlinear recurrence relation which corresponds to the discrete Painlevé I equation and its hierarchy. In Chapter 3 we will study a family of orthogonal polynomials on the unit circle. We will first give some background on orthogonal polynomials on the unit circle and the corresponding recurrence relations. We will study the weight function $w(\theta) = \exp(t \cos \theta)$, and it will be shown that the recurrence coefficients satisfy a nonlinear recurrence relation which corresponds to discrete Painlevé

II. These orthogonal polynomials play an important role in the theory of random unitary matrices and combinatorial problems for random permutations. We will also study certain discrete orthogonal polynomials related to Charlier polynomials. The recurrence coefficients $(a_n)_{n\geq 1}$ and $(b_n)_{n\geq 0}$ are shown to satisfy a system of nonlinear recurrence relations which are again related to the discrete Painlevé II equation. In Chapter 4 we give some details about ladder operators for orthogonal polynomials. These are (differential or difference) operators that map an orthogonal polynomial of degree n to one of degree $n - 1$ (lowering operator) or degree $n + 1$ (raising operator). The compatibility of these ladder operators with the three term recurrence relation gives nonlinear recurrence relations for the recurrence coefficients (a_n, b_n), which can often be identified as discrete Painlevé equations. In Chapter 5 we give some more examples of semi-classical orthogonal polynomials that give rise to discrete and continuous Painlevé equations. In Chapters 6 and 7 we will investigate the six (differential) Painlevé equations but restrict our attention to those aspects that involve orthogonal polynomials. In Chapter 6 we investigate solutions of the Painlevé equations which are in terms of orthogonal polynomials. These are rational solutions and solutions in terms of classical (linear) special functions. Finally, in Chapter 7 we show how Painlevé transcendents appear in the asymptotic analysis of orthogonal polynomials near critical points, i.e., points where the density of the zeros vanishes or becomes unbounded. Usually this corresponds to a phase transition: the zero density is supported on a number of intervals and when these intervals touch or when a new interval appears one often has singular behavior of the zero density.

1.1.1 Pearson equation and semi-classical orthogonal polynomials

Classical orthogonal polynomials are orthogonal with a weight function w on the real line which satisfies a first order differential equation

$$(\sigma w)' = \tau w, \tag{1.13}$$

where σ is a polynomial of degree ≤ 2 and τ a polynomial of degree 1. This equation is known as the *Pearson equation*, named after the statistician Karl Pearson who introduced it in 1895. We are interested in positive solutions w such that σw vanishes at points $a, b \in \mathbb{R} \cup \{-\infty, +\infty\}$. Up to an affine transformation, the classical orthogonal polynomials are

- The *Hermite polynomials*, with $w(x) = e^{-x^2}$ on $(-\infty, \infty)$ and $\sigma = 1$;
- The *Laguerre polynomials*, with $w(x) = x^\alpha e^{-x}$ on $[0, \infty)$ and $\sigma(x) = x$ ($\alpha > -1$);

- The *Jacobi polynomials*, with $w(x) = (1 - x)^{\alpha}(1 + x)^{\beta}$ on $[-1, 1]$ and $\sigma(x) = x^2 - 1$ $(\alpha, \beta > -1)$.

The case $\sigma(x) = x^2$ is related to Bessel polynomials but does not give orthogonal polynomials with a positive measure on the real line. The case $\sigma(x) = x^2 + 1$ gives Romanovski polynomials, but then we can only have a finite number of orthogonal polynomials with a positive measure on the real line, see [101, Thm. 4.1].

Semi-classical orthogonal polynomials have a weight function w that satisfies a Pearson equation (1.13) where σ and τ are polynomials with $\deg \sigma > 2$ or $\deg \tau \neq 1$. We need positive solutions w such that σw vanishes at $a, b \in \mathbb{R} \cup \{-\infty, +\infty\}$. An important property of classical and semi-classical orthogonal polynomials is their *structure relation*:

Property 1.1 *If the weight w satisfies the Pearson equation* (1.13) *and σw vanishes at $a, b \in \mathbb{R} \cup \{-\infty, +\infty\}$, then*

$$\sigma(x)p_n'(x) = \sum_{k=n-t}^{n+s-1} A_{n,k} p_k(x), \tag{1.14}$$

where $s = \deg \sigma$ and $t = \max\{\deg \tau, \deg \sigma - 1\}$.

Proof The polynomial $\sigma p_n'$ has degree $n + s - 1$, so we can expand it in terms of the orthonormal polynomials p_k with $0 \leq k \leq n + s - 1$:

$$\sigma(x)p_n'(x) = \sum_{k=0}^{n+s-1} A_{n,k} p_k(x).$$

The coefficients $A_{n,k}$ are Fourier coefficients and can be expressed as

$$A_{n,k} = \int_a^b \sigma(x)p_n'(x)p_k(x)w(x)\,dx.$$

Integration by parts, and the boundary conditions $\sigma(a)w(a) = 0 = \sigma(b)w(b)$, gives

$$A_{n,k} = -\int_a^b p_n(x)[\sigma(x)w(x)p_k(x)]'\,dx$$

$$= -\int_a^b p_n(x)p_k(x)[\sigma(x)w(x)]'\,dx - \int_a^b p_n(x)p_k'(x)\sigma(x)w(x)\,dx$$

$$= -\int_a^b p_n(x)p_k(x)\tau(x)w(x)\,dx - \int_a^b p_n(x)p_k'(x)\sigma(x)w(x)\,dx,$$

where we used the Pearson equation (1.13) in the last line. By orthogonality the first term vanishes whenever $k + \deg \tau < n$ and the second term vanishes whenever $k + s - 1 < n$, hence both terms vanish whenever $k < n - t$ with $t = \max\{\deg \tau, \deg \sigma - 1\}$ and only the Fourier coefficients $A_{n,k}$ with $n - t \leq k \leq n + s - 1$ are left. $\qquad\qquad\qquad\qquad\qquad\qquad\qquad\qquad\qquad\qquad\square$

Every family of orthogonal polynomials on the real line satisfies a three term recurrence relation (1.2) and semi-classical orthogonal polynomials in addition also satisfy a structure relation (1.14). Both relations should be *compatible*. If we express the compatibility relations in terms of the recurrence coefficients $(a_n)_{n\geq 1}$, $(b_n)_{n\geq 0}$ and the coefficients $(A_{n,k})_{n\geq 1}$ in the structure relation, then we get (nonlinear) recurrence relations for these coefficients. Solving them gives the recurrence coefficients $(a_n)_{n\geq 1}$ and $(b_n)_{n\geq 0}$.

To illustrate this we use the Hermite polynomials, for which $w(x) = e^{-x^2}$ on $(-\infty, +\infty)$ and $\sigma = 1$. The structure relation for the orthonormal Hermite polynomials is

$$p'_n(x) = A_n p_{n-1}(x).$$

Taking derivatives in the three term recurrence relation (1.2) gives

$$p_n(x) + x p'_n(x) = a_{n+1} p'_{n+1}(x) + b_n p'_n(x) + a_n p'_{n-1}(x).$$

Use the structure relation to replace all the derivatives, then

$$p_n(x) + A_n x p_{n-1}(x) = a_{n+1} A_{n+1} p_n(x) + b_n A_n p_{n-1}(x) + a_n A_{n-1} p_{n-2}(x).$$

Now replace $x p_{n-1}(x)$ by using the three term recurrence relation, to find

$$p_n(x) + A_n[a_n p_n(x) + b_{n-1} p_{n-1}(x) + a_{n-1} p_{n-2}(x)]$$
$$= a_{n+1} A_{n+1} p_n(x) + b_n A_n p_{n-1}(x) + a_n A_{n-1} p_{n-2}(x).$$

Since the orthogonal polynomials $\{p_n, p_{n-1}, p_{n-2}\}$ are linearly independent, this expression can only be true if the coefficients in front of p_n, p_{n-1} and p_{n-2} vanish. This gives three equations

$$p_n \Rightarrow 1 + A_n a_n = a_{n+1} A_{n+1}, \qquad (1.15)$$

$$p_{n-1} \Rightarrow A_n b_{n-1} = b_n A_n, \qquad (1.16)$$

$$p_{n-2} \Rightarrow A_n a_{n-1} = a_n A_{n-1}. \qquad (1.17)$$

From (1.16) we find that $b_n = b_{n-1}$ so that b_n is a constant sequence: $b_n = b_0$. From (1.15) we find $a_{n+1} A_{n+1} - a_n A_n = 1$ so that $a_n A_n = n + a_0 A_0$, but we defined $p_{-1} = 0$ so that (1.4) gives $a_0 = 0$. Hence $a_n A_n = n$. Finally (1.17) gives $A_n/a_n = A_{n-1}/a_{n-1}$ so that $A_n/a_n = c$ is constant. Combining this with the previous relation gives $a_n^2 = n/c$ for $n \geq 1$, so that $1/c = a_1^2$. So for Hermite polynomials we were able to solve the nonlinear equations to find

$$b_n = b_0, \qquad a_n^2 = a_1^2 n.$$

We only need to figure out what the initial values b_0 and a_1^2 are, to get all the recurrence coefficients. For b_0 we use (1.5) with $n = 0$ to find

$$b_0 = p_0^2 \int_{-\infty}^{\infty} x e^{-x^2} \, dx = 0,$$

so that $b_n = 0$ for all $n \geq 0$. In fact, this was already clear from the beginning since $w(x) = e^{-x^2}$ is an even weight function. For a_1^2 we use the fact that $p_1(x) = (x - b_0)p_0/a_1$ has norm 1. Recall that

$$p_0^2 = \left(\int_{-\infty}^{\infty} e^{-x^2} \, dx \right)^{-1} = 1/\sqrt{\pi}$$

hence

$$1 = \int_{-\infty}^{\infty} p_1^2(x) e^{-x^2} \, dx = \frac{1}{a_1^2 \sqrt{\pi}} \int_{-\infty}^{\infty} x^2 e^{-x^2} \, dx = \frac{1}{2a_1^2},$$

so that $a_1^2 = 1/2$ and $a_n^2 = n/2$ for $n \geq 1$. Hence the three term recurrence relation for orthonormal Hermite polynomials is

$$\sqrt{2} x p_n(x) = \sqrt{n+1} p_{n+1}(x) + \sqrt{n} p_{n-1}(x)$$

and the structure relation is (recall that $A_n = a_n/a_1^2$)

$$p_n'(x) = \sqrt{2n} p_{n-1}(x).$$

Note that the usual Hermite polynomials $(H_n)_{n \in \mathbb{N}}$ are not orthonormal, but have norm

$$\int_{-\infty}^{\infty} H_n^2(x) e^{-x^2} \, dx = \sqrt{\pi} 2^n n!,$$

hence $H_n(x) = \sqrt{\sqrt{\pi} 2^n n!} \; p_n(x)$. The corresponding recurrence relation and structure relation then become

$$2x H_n(x) = H_{n+1}(x) + 2n H_{n-1}(x), \qquad H_n'(x) = 2n H_{n-1}.$$

Exercise 1: Use the compatibility relations between (1.2) and (1.14) to find the recurrence coefficients for the orthonormal Laguerre polynomials with weight function $w(x) = x^\alpha e^{-x}$ on $[0, \infty)$ for $\alpha > -1$.

1.2 Painlevé equations

1.2.1 The six Painlevé differential equations

Linear differential equations are reasonably easy to investigate. Nonlinear differential equations are a lot harder and several problems, which do not exist for linear equations, appear. One such problem is that the singularities of the solution may depend on the initial conditions. Such singularities are called *movable singularities*. For instance $y' = -y^2$ has the general solution $y(x) = 1/(x - c)$ where c is constant, hence the singularity at $x = c$ depends on the constant of integration, or on the initial value: $c = -1/y(0)$. The movable singularity is a pole in this case. This is not so bad, because it is an isolated singularity. The equation $y' = 1/(2y)$ has the general solution $y(x) = \sqrt{x - c}$ with c a constant. Now the singularities are on a half line in the complex plane starting at c. This is a branch cut and c is a branch point and the location of this branch point depends on the initial condition $c = -y(0)^2$. The movable singularities are not poles but more complicated and depend on the choice of the branch cut. This situation is not desirable and may lead to serious complications when we are comparing solutions of differential equations. Hence, at the end of the 19th century people (Poincaré, Fuchs, Picard, Painlevé) became interested in finding those nonlinear differential equations for which **the general solution is free from movable branch points**. This is called the *Painlevé property*. The locations of possible branch points and critical essential singularities of solutions may not depend on the initial values. For first order differential equations the Painlevé property only gives linear differential equations, the Weierstrass elliptic function $\wp$ satisfying $(y')^2 = 4y^3 - g_2 y - g_3$ or the Riccati differential equation $y' = q_0(x) + q_1(x)y + q_2(x)y^2$. Picard raised the problem of finding the nonlinear differential equations of the form $y'' = R(y', y, x)$, where R is a rational function, with the Painlevé property. At the beginning of the 20th century Paul Painlevé found that, up to certain simple transformations, these differential equations can be put into one of 50 canonical forms. Out of these 50, there are 44 that can be reduced to linear equations, the Weierstrass elliptic equation, the Riccati equation, or one of six equations of the list. These six equations are now known as the Painlevé equations and their solutions are called *Painlevé transcendents*. It turns out that for these second order equations the only movable singularities are poles (no essential singularities). These Painlevé equations are important nonlinear special functions that nowadays appear in integrable systems, statistical mechanics, random matrix theory and orthogonal polynomials. The six equations are

$$P_I \quad y'' = 6y^2 + x, \tag{1.18}$$

$$P_{II} \quad y'' = 2y^3 + xy + \alpha, \tag{1.19}$$

$$P_{III} \quad y'' = \frac{(y')^2}{y} - \frac{y'}{x} + \frac{\alpha y^2 + \beta}{x} + \gamma y^3 + \frac{\delta}{y}, \tag{1.20}$$

$$P_{IV} \quad y'' = \frac{(y')^2}{2y} + \frac{3}{2}y^3 + 4xy^2 + 2(x^2 - \alpha)y + \frac{\beta}{y}, \tag{1.21}$$

$$P_V \quad y'' = \left(\frac{1}{2y} + \frac{1}{y-1}\right)(y')^2 - \frac{y'}{x} + \frac{(y-1)^2}{x^2}\left(\alpha y + \frac{\beta}{y}\right) + \frac{\gamma y}{x}$$
$$+ \frac{\delta y(y+1)}{y-1}, \tag{1.22}$$

$$P_{VI} \quad y'' = \frac{1}{2}\left(\frac{1}{y} + \frac{1}{y-1} + \frac{1}{y-x}\right)(y')^2 - \left(\frac{1}{x} + \frac{1}{x-1} + \frac{1}{y-x}\right)y'$$
$$+ \frac{y(y-1)(y-x)}{x^2(x-1)^2}\left(\alpha + \frac{\beta x}{y^2} + \frac{\gamma(x-1)}{(y-1)^2} + \frac{\delta x(x-1)}{(y-x)^2}\right), \tag{1.23}$$

where $\alpha, \beta, \gamma, \delta$ are constants. A good survey can be found in [39].

1.2.2 Discrete Painlevé equations

Discrete Painlevé equations appeared more recently. They are nonlinear "integrable" discrete equations (recurrence relations) for which the continuous limit is one of the Painlevé differential equations. Usually only second order equations are considered. The term *integrable* remains ambiguous: what do we mean by that? Without giving a precise meaning it basically means that anything simpler becomes linear, anything more complicated becomes hopelessly complicated. For a good survey we refer to [80].

There is a discrete version of the Painlevé property which one can use as a detector for integrability. This notion is *singularity confinement*. Suppose that we are dealing with a recurrence relation $x_n = f(x_{n-2}, x_{n-1}, n)$, with f a rational function. Let n_0 be an index such that $(x_{n_0-2}, x_{n_0-1}, n_0)$ gives a singularity for f, so that x_{n_0} is not defined. Then singularity confinement means that there is an integer p such that the singularity is confined to the elements $x_{n_0}, x_{n_0+1}, \ldots, x_{n_0+p}$ but x_{n_0+p+1} is again defined and it depends on what happened before the singularities, i.e., on x_{n_0-1}. So the singularity is restricted to a finite section of the sequence $(x_n)_{n \in \mathbb{N}}$, which is the discrete version of an isolated singularity (a pole) for complex functions. To check for singularity confinement, one usually starts from $\tilde{x}_{n_0-1} = x_{n_0-1} + \epsilon$ and one computes

$$\tilde{x}_{n_0} = O(\frac{1}{\epsilon}) + \cdots$$

$$\vdots$$

$$\tilde{x}_{n_0+p} = O(\frac{1}{\epsilon}) + \cdots$$

$$\tilde{x}_{n_0+p+1} = x_{n_0+p+1} + O(\epsilon)$$

with a careful analysis of the error terms. Then as $\epsilon \to 0$ we can find the value of p and we can see how x_{n_0+p+1} depends on the past (before the singularity). The property of singularity confinement however *does not* characterize discrete Painlevé equations: there are examples of discrete equations with singularity confinement, which we should not call a discrete Painlevé equation. For this reason, singularity confined is only used as a discrete integrability detector.

Making a canonical list of discrete Painlevé equations is more complicated than for differential equations since we cannot use transformations of the variable n to construct an equivalence class of equations of the same type. We still can use transformations of the solution, of course. A list of standard discrete Painlevé equations grew historically as the equations appeared. A partial list is

$$\text{d-P}_\text{I} \quad x_{n+1} + x_n + x_{n-1} = \frac{z_n + a(-1)^n}{x_n} + b, \tag{1.24}$$

$$\text{d-P}_\text{II} \quad x_{n+1} + x_{n-1} = \frac{x_n z_n + a}{1 - x_n^2}, \tag{1.25}$$

$$\text{d-P}_\text{IV} \quad (x_{n+1} + x_n)(x_n + x_{n-1}) = \frac{(x_n^2 - a^2)(x_n^2 - b^2)}{(x_n + z_n)^2 - c^2}, \tag{1.26}$$

$$\text{d-P}_\text{V} \quad \frac{(x_{n+1} + x_n - z_{n+1} - z_n)(x_n + x_{n-1} - z_n - z_{n-1})}{(x_{n+1} + x_n)(x_n + x_{n-1})}$$

$$= \frac{[(x_n - z_n)^2 - a^2][(x_n - z_n)^2 - b^2]}{(x_n - c^2)(x_n - d^2)}, \tag{1.27}$$

where $z_n = \alpha n + \beta$ and a, b, c, d are constants. Observe that there is no d-P$_\text{III}$ or d-P$_\text{VI}$. That is because in the above equations the x_n and x_{n+1} (and x_n and x_{n-1}) appear in an additive way. There are other discrete Painlevé equations where they appear in a multiplicative way:

$$\text{q-P}_{\text{III}} \quad x_{n+1}x_{n-1} = \frac{(x_n - aq_n)(x_n - bq_n)}{(1 - cx_n)(1 - x_n/c)}, \tag{1.28}$$

$$\text{q-P}_{\text{V}} \quad (x_{n+1}x_n - 1)(x_n x_{n-1} - 1) = \frac{(x_n - a)(x_n - 1/a)(x_n - b)(x_n - 1/b)}{(1 - cx_n q_n)(1 - x_n q_n/c)}, \tag{1.29}$$

$$\text{q-P}_{\text{VI}} \quad \frac{(x_n x_{n+1} - q_n q_{n+1})(x_n x_{n-1} - q_n q_{n-1})}{(x_n x_{n+1} - 1)(x_n x_{n-1} - 1)}$$

$$= \frac{(x_n - aq_n)(x_n - q_n/a)(x_n - bq_n)(x_n - q_n/b)}{(x_n - c)(x_n - 1/c)(x_n - d)(x_n - 1/d)}, \tag{1.30}$$

where $q_n = q_0 q^n$ and a, b, c, d are constants. There are also various asymmetric discrete Painlevé equations which are given as a system of two first order equations for the unknowns x_n and y_n. Just to give one example

$$\alpha\text{-d-P}_{\text{IV}} \quad (x_n + y_n)(x_{n+1} + y_n) = \frac{(y_n - a)(y_n - b)(y_n - c)(y_n - d)}{(y_n + \gamma - z_n)(y_n - \gamma - z_n)}$$

$$(x_n + y_n)(x_n + y_{n-1}) = \frac{(x_n + a)(x_n + b)(x_n + c)(x_n + d)}{(x_n + \delta - z_{n+1/2})(x_n - \delta - z_{n+1/2})}. \tag{1.31}$$

In 2001 H. Sakai suggested a classification of discrete Painlevé equations based on *rational surfaces* associated with affine root systems, starting from root system E_8. Sakai obtained discrete Painlevé equations corresponding to the following degeneracy pattern, where we indicate the relevant root system:

$$
\begin{array}{ccccccccccccc}
E_8^e & & & & & & & & & & & & A_1^q \\
\downarrow & & & & & & & & & & & \nearrow & \\
E_8^q & \to & E_7^q & \to & E_6^q & \to & D_5^q & \to & A_4^q & \to & (A_2 + A_1)^q & \to & (A_1 + A_1)^q & \to & A_1^q \\
\downarrow & & \downarrow & & \downarrow & & \downarrow & & \downarrow & & | \quad \downarrow & & | \quad \downarrow & \\
E_8^d & \to & E_7^d & \to & E_6^d & \to & D_4^c & \to & A_3^c & \to & | \quad (2A_1)^c & \to & | \quad A_1^c & \\
& & & & & & & \searrow & & \downarrow & & \searrow & \downarrow \\
& & & & & & & & & A_2^c & & \to & A_1^c &
\end{array}
$$

The diagram contains the exceptional root systems E_8, E_7, E_6 and the root systems D_4, D_5 and A_1, A_2, A_3, A_4. The superscript e indicates a discrete system involving elliptic functions, the superscript q denotes an equation of type q-P, the superscript d an equation of type d-P and the superscript c denotes a difference equation which is a contiguity relation of one of the continuous Painlevé equations. In this setting our α-d-P$_{\text{IV}}$ corresponds to E_6^d.

Exercise 2: Show that the *continuum limit* of d-P$_{\text{I}}$ in (1.24), with $x_n = 1 - 2h^2 y(x)$, $\alpha = 2h^5$, $\beta = -3$, $a = 0$ and $b = 6$, where $x = nh$, is Painlevé I given in (1.18) when $h \to 0$, with a change of sign in x.

Bonus: For which continuum limit will d-P$_{\text{II}}$ in (1.25) be Painlevé II in (1.19)? (Hint: try $x_n = hy(x)$ with $x = nh$ and α, β, a depending on h.)

Sakai's work is based on Okamoto's work [132], which is the construction of the space of initial values which parametrizes all the solutions for each Painlevé equation. Geometrically, the space of initial values is a surface which is characterized by a pair of affine root systems which represent the *symmetry type* and the *surface type*. Kajiwara, Noumi and Yamada [96, §3.4] gave credit to Okamoto for observing a remarkable complementary relation between the surface type and symmetry type. Their paper [96] is strongly recommended for those that want to understand the geometric aspects of Painlevé equations. In that paper, Kajiwara, Noumi and Yamada propose to refer to the (discrete) Painlevé equations by using both their surface type and symmetry type. They give a list of basic data for the discrete Painlevé equations in [96, §8]. In their terminology our α-d-P$_{IV}$ in (1.31) corresponds to d-P$(E_6^{(1)}/A_2^{(1)})$, where $E_6^{(1)}$ is the surface type, $A_2^{(1)}$ is the symmetry type and the superscript (1) indicates that we are dealing with the affine Weyl groups corresponding to the root systems E_6 and A_2.

2

Freud weights and discrete Painlevé I

Freud weights are exponential weights on the real line $(-\infty, +\infty)$ of the form

$$w(x) = |x|^\rho \exp(-|x|^m), \qquad \rho > -1, m > 0.$$

They were introduced by Géza Freud in 1976 [77] and he gave a recurrence relation for the recurrence coefficients $(a_n)_{n \geq 1}$ for the case $m = 2, 4, 6$. Later it was realized that Freud's recurrence relations are discrete Painlevé equations (and their higher order versions) and it turned out that Shohat already had some of these relations in 1939 [142, Eq. (39) and (42)] and Laguerre [106] in 1885. Laguerre's work was related to the continued fraction coefficients for a function satisfying a differential equation like Pearson's equation (1.13), so one needs the relation between the continued fraction coefficients and the recurrence coefficients in (1.2) to translate Laguerre's results to the present setting.

2.1 The Freud weight $w(x) = e^{-x^4 + tx^2}$

We will consider the exponential weight

$$w(x) = e^{-x^4 + tx^2}, \qquad x \in (-\infty, +\infty), \tag{2.1}$$

where $t \in \mathbb{R}$ is a parameter. The polynomials in the Pearson equation (1.13) are $\sigma = 1$ and $\tau(x) = -4x^3 + 2tx$. This is a symmetric weight function, so that the recurrence relation (1.2) for the orthonormal polynomials is

$$xp_n(x) = a_{n+1}p_{n+1}(x) + a_n p_{n-1}(x), \tag{2.2}$$

and the structure relation (1.14) simplifies to

$$p'_n(x) = A_n p_{n-1}(x) + C_n p_{n-3}(x). \tag{2.3}$$

13

The compatibility relations now lead to the following result, which for $t = 0$ was obtained by Shohat [142] and Freud [77].

Theorem 2.1   *The recurrence coefficients $(a_n)_{n \geq 1}$ of the orthonormal polynomials with the weight (2.1) satisfy*

$$4a_n^2 \left(a_{n+1}^2 + a_n^2 + a_{n-1}^2 - \frac{t}{2} \right) = n, \tag{2.4}$$

with initial conditions $a_0 = 0$ and

$$a_1^2 = \frac{\int_{-\infty}^{\infty} x^2 e^{-x^4 + tx^2} \, dx}{\int_{-\infty}^{\infty} e^{-x^4 + tx^2} \, dx}.$$

Observe that (2.4) is the discrete Painlevé I equation d-P$_{\text{I}}$ in (1.24) with $x_n = a_n^2$ and parameters $a = 0$, $b = t/2$, $\alpha = 1/4$ and $\beta = 0$.

Proof   If we take derivatives in the three term recurrence relation (2.2) and then replace all the derivatives using the structure relation (2.3), then one finds

$$p_n(x) + x(A_n p_{n-1}(x) + C_n p_{n-3}(x))$$
$$= a_{n+1}(A_{n+1} p_n(x) + C_{n+1} p_{n-2}(x)) + a_n(A_{n-1} p_{n-2}(x) + C_{n-1} p_{n-4}(x)).$$

Now replace $x p_{n-1}(x)$ and $x p_{n-3}(x)$ by using the three term recurrence relation (2.2), to find

$$p_n(x) + A_n(a_n p_n(x) + a_{n-1} p_{n-2}(x)) + C_n(a_{n-2} p_{n-2}(x) + a_{n-3} p_{n-4}(x))$$
$$= a_{n+1}(A_{n+1} p_n(x) + C_{n+1} p_{n-2}(x)) + a_n(A_{n-1} p_{n-2}(x) + C_{n-1} p_{n-4}(x)).$$

Comparing the coefficients of p_n, p_{n-2}, p_{n-3} gives

$$p_n \Rightarrow 1 + A_n a_n = a_{n+1} A_{n+1}, \tag{2.5}$$
$$p_{n-2} \Rightarrow A_n a_{n-1} + C_n a_{n-2} = a_{n+1} C_{n+1} + a_n A_{n-1}, \tag{2.6}$$
$$p_{n-4} \Rightarrow C_n a_{n-3} = a_n C_{n-1}. \tag{2.7}$$

From (2.5) we find that $a_{n+1} A_{n+1} - a_n A_n = 1$ so that

$$a_n A_n = n, \qquad n \geq 1. \tag{2.8}$$

Here we used $a_0 = 0$. From (2.7) we find that

$$\frac{C_n}{a_n a_{n-1} a_{n-2}} = \frac{C_{n-1}}{a_{n-1} a_{n-2} a_{n-3}},$$

so that $C_n / (a_n a_{n-1} a_{n-2})$ is constant, hence

$$C_n = c a_n a_{n-1} a_{n-2}, \qquad n \geq 3, \tag{2.9}$$

with c a constant. If we use (2.8)–(2.9) in (2.6), then we find

$$na_{n-1}^2 - (n-1)a_n^2 = ca_n^2 a_{n-1}^2 (a_{n+1}^2 - a_{n-2}^2).$$

Divide both sides by $a_n^2 a_{n-1}^2$ to get

$$\frac{n}{a_n^2} - \frac{n-1}{a_{n-1}^2} = c(a_{n+1}^2 - a_{n-2}^2).$$

Summing from 2 to n gives

$$\frac{n}{a_n^2} - \frac{1}{a_1^2} = c(a_{n+1}^2 + a_n^2 + a_{n-1}^2 - a_2^2 - a_1^2)$$

and multiplying by a_n^2 then leads to

$$n = a_n^2/a_1^2 + ca_n^2(a_{n+1}^2 + a_n^2 + a_{n-1}^2) - ca_n^2(a_1^2 + a_2^2). \tag{2.10}$$

This equation still contains the constant c and the initial values a_1^2 and a_2^2, which we need to determine. To find c, we use the structure relation (2.3) and integrate:

$$C_3 = \int_{-\infty}^{\infty} p_3'(x)p_0 e^{-x^4 + tx^2} \, dx.$$

Integration by parts gives

$$C_3 = -p_0 \int_{-\infty}^{\infty} p_3(x)(-4x^3 + 2tx)e^{-x^4 + tx^2} \, dx = 4p_0 \int_{-\infty}^{\infty} p_3(x)x^3 e^{-x^4 + tx^2} \, dx,$$

where we used the orthogonality in the last step. Now use

$$p_3(x) = \gamma_3 x^3 + \cdots, \qquad \gamma_3 = \frac{\gamma_0}{a_1 a_2 a_3},$$

where we used (1.3) to express γ_n in terms of the recurrence coefficients $(a_n)_{n \geq 1}$, then $C_3 = 4a_1 a_2 a_3$ so that $c = 4$.

Next, we introduce the moments

$$m_k = \int_{-\infty}^{\infty} x^k e^{-x^4 + tx^2} \, dx,$$

then $m_k = 0$ whenever k is odd. Integration by parts gives

$$m_0 = -\int_{-\infty}^{\infty} x(-4x^3 + 2tx)e^{-x^4 + tx^2} \, dx = 4m_4 - 2tm_2,$$

so that $m_4 = (m_0 + 2tm_2)/4$. We have $p_0^2 = 1/m_0$ and $p_1(x) = xp_0/a_1$, hence the orthonormality of p_1 gives $a_1^2 = m_2/m_0$, which is the initial condition given in the theorem. Observe that

$$a_1^2 + a_2^2 = \int_{-\infty}^{\infty} [xp_1(x)]^2 e^{-x^4 + tx^2} \, dx$$

(Parseval's identity), hence

$$a_1^2 + a_2^2 = \frac{p_0^2}{a_1^2} m_4 = \frac{1}{4} \frac{m_0 + 2tm_2}{m_2} = \frac{1}{4}\left(\frac{1}{a_1^2} + 2t\right),$$

hence (2.10) becomes the required equation (2.4). □

Exercise 3: Show that the recurrence coefficients for the weights $w(x) = |x|^\rho e^{-x^4}$ satisfy

$$4a_n^2(a_{n+1}^2 + a_n^2 + a_{n-1}^2) = n + \rho\Delta_n$$

where

$$\Delta_n = \begin{cases} 0 & \text{if } n \text{ is even,} \\ 1 & \text{if } n \text{ is odd.} \end{cases}$$

Hint: this weight satisfies a Pearson equation with $\sigma(x) = x$.

2.2 Asymptotic behavior of the recurrence coefficients

We can find the asymptotic behavior of the recurrence coefficients from the recurrence relation (2.4).

Property 2.2 (Freud)   *The recurrence coefficients for the weight* $w(x) = e^{-x^4 + tx^2}$ *satisfy*

$$\lim_{n\to\infty} \frac{a_n}{n^{1/4}} = \frac{1}{\sqrt[4]{12}}.$$

Proof   If one knows that the limit exists, then it is easy to find the limit by taking limits in the relation (2.4) (after dividing by n). So the difficulty is to show that the limit exists. First we look for an upper bound. From (2.4) we find

$$4a_n^4 + 4a_n^2(a_{n+1}^2 + a_{n-1}^2) - 2ta_n^2 = n$$

so that $4a_n^4 - 2ta_n^2 \leq n$, and hence $(2a_n^2 - t/2)^2 \leq n + t^2/4$. Hence $a_n^2/\sqrt{n}$ is bounded and positive. Let

$$A = \liminf_{n\to\infty} \frac{a_n^2}{\sqrt{n}}, \qquad B = \limsup_{n\to\infty} \frac{a_n^2}{\sqrt{n}},$$

then clearly $A \leq B$. Let $(a_{n_k})_{k\in\mathbb{N}}$ be a subsequence such that

$$A = \lim_{k\to\infty} \frac{a_{n_k}^2}{\sqrt{n_k}},$$

then taking limits over this subsequence in (2.4) gives $1 \leq 4A(2B + A)$. In a similar way we can take a subsequence $(a_{m_k})_{k\in\mathbb{N}}$ such that

$$B = \lim_{k\to\infty} \frac{a_{m_k}^2}{\sqrt{m_k}},$$

then taking limits over this subsequence in (2.4) gives $1 \geq 4B(2A + B)$. Combining both inequalities gives $4B(2A + B) \leq 4A(2B + A)$, which gives $B^2 \leq A^2$, but that means that $A = B$ and the limit indeed exists. Clearly $A^2 = 1/12$, from which the result follows. $\qquad\square$

Freud [77] gave the asymptotic behavior of the recurrence coefficients for the weights $w(x) = |x|^\rho \exp(-|x|^m)$ when $m = 2, 4, 6$ and he formulated a conjecture for general $m > 0$, which was later proved by Magnus [111] in 1984 for even integers m and Lubinsky, Mhaskar and Saff in [108] 1986 for $m > 0$. An asymptotic expansion for the recurrence coefficients for the weight e^{-x^4} was obtained by Máté, Nevai and Zaslavsky [119] in 1985. These were considered as important results in the theory of orthogonal polynomials, and they were obtained before people realized that one was dealing with a discrete Painlevé equation, which was first noticed by Fokas, Its and Kitaev [69] in 1991.

2.3 Unicity of the positive solution of d-P_I with $x_0 = 0$

The recurrence relation (2.4) is not so good for computing the recurrence coefficients recursively, starting from $a_0^2 = 0$ and $a_1^2 = m_2/m_0$. First of all the integrals m_0 and m_2 are difficult to compute: they depend on the modified Bessel function $K_{1/4}(t^2/8)$ or the parabolic cylinder function $D_{-1/2}(t/\sqrt{2})$, and for $t = 0$ on the Gamma functions $\Gamma(3/4)$ and $\Gamma(1/4)$. Worst of all is that the recurrence relation is very sensitive to errors in the initial value. This is a property that often occurs in nonlinear systems: a small change in the initial value may lead to big changes in the solution.

The reason for this *butterfly effect* is that the solution that we want is a positive solution with $a_0^2 = 0$ and this solution turns out to be unique, i.e., there is only one initial value for a_1^2 that leads to a solution which is positive for every $n \geq 0$. Hence a slight error in computing $a_1^2 = m_2/m_0$ will eventually give a coefficient a_n^2 which is not positive (see Figure 2.1). We will give a proof of this result for the case $t = 0$.

Theorem 2.3 (Lew and Quarles [107], Nevai [122])   *There is a unique solution of*

$$x_n(x_{n+1} + x_n + x_{n-1}) = an, \qquad a > 0, \tag{2.11}$$

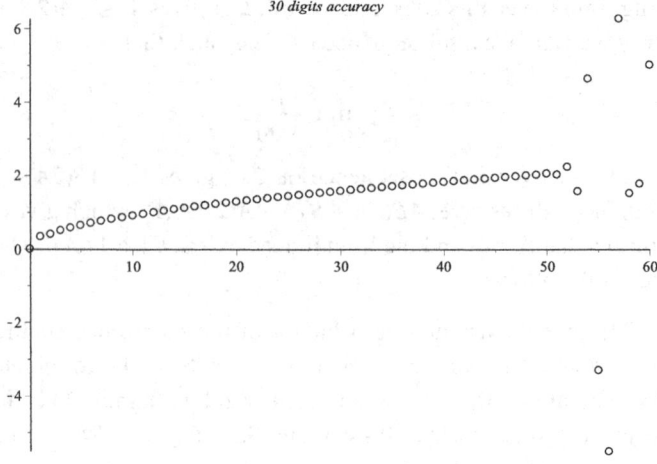

Figure 2.1 Computing a_n^2 from (2.4) (with $t = 0$) using 30 significant digits

for which $x_0 = 0$ and $x_n > 0$ for all $n \geq 1$. This positive solution corresponds to the initial value $x_1 = \sqrt{4a}\Gamma(3/4)/\Gamma(1/4)$.

Proof   If we solve the quadratic equation (2.11) for x_n, then

$$x_n = \frac{-(x_{n+1} + x_{n-1}) \pm \sqrt{(x_{n+1} + x_{n-1})^2 + 4an}}{2}.$$

We are looking for a positive solution, hence we choose the positive sign before the square root. Hence the positive solutions of (2.11) satisfy

$$x_n = \sqrt{an}\, g\left(\frac{x_{n+1} + x_{n-1}}{2\sqrt{an}}\right), \qquad n \geq 1,$$

where $g : [0, \infty) \to [0, 1]$ is given by

$$g(x) = \sqrt{1 + x^2} - x = \frac{1}{x + \sqrt{1 + x^2}}. \tag{2.12}$$

Observe that $g'(x) = -g(x)/\sqrt{1 + x^2}$ so that g is a positive and decreasing function with $g(0) = 1$, $g'(0) = -1$ and $\lim_{x \to \infty} g(x) = 0$. Consider the linear space

$$\mathbb{R}^{\mathbb{N}_0} = \{(x_n)_{n \geq 0} \mid x_0 = 0 \text{ and } x_n \in \mathbb{R} \text{ for all } n \geq 1\},$$

with norm

$$\|x\| = \sup_{n \geq 1} \frac{|x_n|}{\sqrt{n}}.$$

In $\mathbb{R}^{\mathbb{N}_0}$ we take the sets

$$B_{s,r} = \{(x_n)_{n\geq 0} \mid \forall n \geq 0 : s\sqrt{n} \leq x_n \leq r\sqrt{n}\}, \qquad 0 \leq s \leq r,$$

then these sets are complete for this norm. We now introduce a mapping T on infinite sequences $T : \mathbb{R}^{\mathbb{N}_0} \to \mathbb{R}^{\mathbb{N}_0}$, such that $x \mapsto Tx$ with

$$(Tx)_n = \begin{cases} 0, & \text{if } n = 0, \\ \sqrt{an}\, g\left(\frac{x_{n+1}+x_{n-1}}{2\sqrt{an}}\right), & \text{if } n \geq 1, \end{cases}$$

where g is given in (2.12). The positive solution of (2.11) with $x_0 = 0$ is a fixed point of this mapping: $x = Tx$. We will show that there is a unique fixed point in the set $B_{g(1)\sqrt{a},\sqrt{a}}$, which will prove the theorem.

We introduce a partial order relation in $\mathbb{R}^{\mathbb{N}_0}$ by saying that $x \leq y$ whenever $x_n \leq y_n$ for every $n \geq 1$. If we denote by $\mathbf{0}$ the sequence $(0,0,0,\ldots)$, then $T\mathbf{0} = (\sqrt{an})_{n\geq 0} \in B_{s,r}$ whenever $s \leq \sqrt{a} \leq r$. We also have

$$(T^2\mathbf{0})_n = \sqrt{an}\, g\left(\frac{\sqrt{n+1}+\sqrt{n-1}}{2\sqrt{n}}\right), \qquad n \geq 1,$$

and since

$$\sqrt{2} \leq \frac{\sqrt{n+1}+\sqrt{n-1}}{\sqrt{n}} \leq 2, \qquad n \geq 1, \tag{2.13}$$

we see that $T^2\mathbf{0} \in B_{g(1)\sqrt{a},g(1/\sqrt{2})\sqrt{a}} \subset B_{g(1)\sqrt{a},\sqrt{a}}$. Furthermore, since g is decreasing, we have that $x \leq y$ implies $Tx \geq Ty$ and $T^2x \leq T^2y$. Let $\mathbf{0} \leq x$ then $\mathbf{0} \leq Tx \leq T\mathbf{0}$, where the first inequality follows since g is a positive function, hence $Tx \in B_{0,\sqrt{a}}$. If $\mathbf{0} \leq x \leq T\mathbf{0}$ then $T^2\mathbf{0} \leq Tx \leq T\mathbf{0}$, hence $Tx \in B_{g(1)\sqrt{a},\sqrt{a}}$. We conclude that for each x with $\mathbf{0} \leq x$ we have that $T^2x \in B_{g(1)\sqrt{a},\sqrt{a}}$.

If $x, y \in B_{g(1)\sqrt{a},\sqrt{a}}$ then for $n \geq 1$

$$(Tx)_n - (Ty)_n = \sqrt{an}\left(g\left(\frac{x_{n+1}+x_{n-1}}{2\sqrt{an}}\right) - g\left(\frac{y_{n+1}+y_{n-1}}{2\sqrt{an}}\right)\right),$$

so that

$$\frac{|(Tx)_n - (Ty)_n|}{\sqrt{n}} = \sqrt{a}\left|g\left(\frac{x_{n+1}+x_{n-1}}{2\sqrt{an}}\right) - g\left(\frac{y_{n+1}+y_{n-1}}{2\sqrt{an}}\right)\right|$$

$$= \sqrt{a}|g'(\xi_n)|\left|\frac{x_{n+1}+x_{n-1}}{2\sqrt{an}} - \frac{y_{n+1}+y_{n-1}}{2\sqrt{an}}\right|$$

$$\leq \frac{|g'(\xi_n)|}{2}\left(\frac{\sqrt{n+1}}{\sqrt{n}}\frac{|x_{n+1}-y_{n+1}|}{\sqrt{n+1}} + \frac{\sqrt{n-1}}{\sqrt{n}}\frac{|x_{n-1}-y_{n-1}|}{\sqrt{n-1}}\right),$$

where we used Lagrange's mean value theorem, and ξ_n is between $(x_{n+1} +$

$x_{n-1})/2 \sqrt{an}$ and $(y_{n+1} + y_{n-1})/2 \sqrt{an}$. Since $x, y \in B_{g(1) \sqrt{a}, \sqrt{a}}$ we have that $x_n \geq g(1) \sqrt{an}$ and $y_n \geq g(1) \sqrt{an}$, hence

$$\frac{x_{n+1} + x_{n-1}}{2 \sqrt{an}} \geq \frac{g(1)}{2} \frac{\sqrt{n+1} + \sqrt{n-1}}{\sqrt{n}} \geq \frac{g(1)}{\sqrt{2}},$$

where we used (2.13). A similar lower bound holds for y. Hence $\xi_n \geq g(1)/ \sqrt{2}$ so that $|g'(\xi_n)| \leq |g'(g(1)/ \sqrt{2})| := c < 1$. We therefore have

$$\frac{|(Tx)_n - (Ty)_n|}{\sqrt{n}} \leq \frac{c}{2} \frac{\sqrt{n+1} + \sqrt{n-1}}{\sqrt{n}} \|x - y\| \leq c \|x - y\|,$$

where we used the upper bound in (2.13). We then conclude that $\|Tx - Ty\| \leq c\|x - y\|$ when $x, y \in B_{g(1) \sqrt{a}, \sqrt{a}}$. Hence T is a contraction on $B_{g(1) \sqrt{a}, \sqrt{a}}$. If we start with a positive sequence $\mathbf{0} \leq x$, then after applying T twice we are in $B_{g(1) \sqrt{a}, \sqrt{a}}$. Hence the contraction theorem (Banach's fixed point theorem) tells us that there is a unique fixed point in the set $B_{g(1) \sqrt{a}, \sqrt{a}}$ and this corresponds to the positive solution of (2.11) with $x_0 = 0$. The fixed point can be found as $\lim_{k \to \infty} T^k x$ for any x for which $\mathbf{0} \leq x$.

This positive solution corresponds to $x_n = \sqrt{4a} a_n^2$, where $(a_n)_n$ are the recurrence coefficients for the orthogonal polynomials with weight $w(x) = e^{-x^4}$. The first coefficient a_1^2 is given by the ratio m_2/m_0 for the first two non-zero moments, and this gives $a_1^2 = \Gamma(3/4)/\Gamma(1/4)$. $\square$

The case with $t \neq 0$ was first considered by Bonan and Nevai [17]. See also [2] for more details on the history of this result and for a more general theorem. The proof of this theorem actually gives a stable way to compute this positive solution by using an iterative procedure, for which the fixed point gives the desired solution. We can start from the initial sequence $\mathbf{0}$, then $\mathbf{0} \leq T^2 \mathbf{0}$, which implies

$$T^{2k} \mathbf{0} \leq T^{2k+2} \mathbf{0}, \qquad T^{2k+1} \mathbf{0} \leq T^{2k-1} \mathbf{0},$$

so that $T^{2k} \mathbf{0}$ is increasing and $T^{2k+1} \mathbf{0}$ is decreasing. From $\mathbf{0} \leq T \mathbf{0}$ we find $T^{2k} \mathbf{0} \leq T^{2k+1} \mathbf{0}$, so that

$$\mathbf{0} \leq T^2 \mathbf{0} \leq T^{2k} \mathbf{0} \leq T^{2k+1} \mathbf{0} \leq T \mathbf{0}, \qquad k \geq 1.$$

The sequence $T^k \mathbf{0}$ converges to the fixed point x^* and we can control the process since $T^{2k} \mathbf{0} \leq x^* \leq T^{2k+1} \mathbf{0}$. Of course we can only compute a finite number of terms, so in order to compute n terms $(x_1^*, \ldots, x_n^*)$ of the fixed point x^* we need to decide how many iterations we are going to use (say N), and then start with $N + n$ terms of the initial sequence, since each time we use T we will lose one term, due to the fact that for $(Tx)_n$ we need x_{n+1}. In MATLAB$^{\circledR}$ this would be

```
x=[0,zeros(1,n+N)];
for k=1:N
    y(1)=0;
    for j=2:n+N-k+1
        y(j)=(-(x(j+1)+x(j-1))
                +sqrt(4*(j-1)+(x(j+1)+x(j-1))^2))/2;
    end
    x=y(1:n+N-k+1);
end
```

Here N is the number of iterations we use to compute $(x_1, x_2, \ldots, x_n)$. Note however that we have shifted the index by one to avoid using $x(0)$ or $y(0)$. This iteration process does not require computations in high precision, so there is no need for software with high precision routines.

2.4 The Langmuir lattice

We will show that the recurrence coefficients of the orthogonal polynomials for the weight $w(x) = e^{-x^4 + tx^2}$ also satisfy a Painlevé differential equation in the variable t. In order to do that, we first give a differential-difference equation for the recurrence coefficients of orthogonal polynomials with a symmetric weight that is multiplied by an exponential function.

Theorem 2.4 *Let μ be a symmetric positive measure on the real line for which all the moments exist and let μ_t be the measure for which $d\mu_t(x) = e^{tx^2} d\mu(x)$, where $t \in \mathbb{R}$ is such that all the moments of μ_t exist[1]. Then the recurrence coefficients of the orthogonal polynomials for the measure μ_t satisfy the differential-difference equation*

$$\frac{d}{dt} a_n^2 = a_n^2 \left(a_{n+1}^2 - a_{n-1}^2 \right), \qquad n \geq 1. \tag{2.14}$$

Proof Since μ and μ_t are symmetric measures on the real line, the three term recurrence relation for the orthogonal polynomials is

$$x p_n(x; t) = a_{n+1}(t) p_{n+1}(x; t) + a_n(t) p_{n-1}(x; t).$$

We have expressed the dependence on t explicitly, but in the remainder we will drop this dependence. Taking derivatives with respect to t in the recurrence

---

[1] Every $t \leq 0$ automatically satisfies this assumption.

relation gives

$$x\frac{d}{dt}p_n(x) = \left(\frac{d}{dt}a_{n+1}\right)p_{n+1}(x) + a_{n+1}\frac{d}{dt}p_{n+1}(x) + \left(\frac{d}{dt}a_n\right)p_{n-1}(x) + a_n\frac{d}{dt}p_{n-1}(x).$$

Multiply this by $p_{n+1}(x)$ and integrate using the measure μ_t, to find

$$\int x\left(\frac{d}{dt}p_n(x)\right)p_{n+1}(x)\,d\mu_t(x) = \frac{d}{dt}a_{n+1} + a_{n+1}\int\left(\frac{d}{dt}p_{n+1}(x)\right)p_{n+1}(x)\,d\mu_t(x),$$

where we used the orthonormality of the polynomials $(p_n)_{n\in\mathbb{N}}$. In the integral on the left-hand side we use the three term recurrence relation for $xp_{n+1}(x)$ to find

$$\frac{d}{dt}a_{n+1} = a_{n+1}\left(\int\left(\frac{d}{dt}p_n(x)\right)p_n(x)\,d\mu_t(x) - \int\left(\frac{d}{dt}p_{n+1}(x)\right)p_{n+1}(x)\,d\mu_t(x)\right).$$

$$(2.15)$$

The orthonormality gives

$$\int p_n^2(x)e^{tx^2}\,d\mu(x) = 1,$$

and by differentiating with respect to t we have

$$2\int\left(\frac{d}{dt}p_n(x)\right)p_n(x)\,d\mu_t(x) + \int x^2p_n^2(x)\,d\mu_t(x) = 0,$$

which gives

$$\int\left(\frac{d}{dt}p_n(x)\right)p_n(x)\,d\mu_t(x) = -\frac{1}{2}(a_{n+1}^2 + a_n^2).$$

Inserting in (2.15) then gives

$$\frac{d}{dt}a_{n+1} = \frac{1}{2}\,a_{n+1}\left(a_{n+2}^2 - a_n^2\right),$$

which is the desired equation for $n + 1$, since $\frac{d}{dt}a_n^2 = 2a_n\frac{d}{dt}a_n$.   □

The system of equations $x_n'(t) = x_n(t)(x_{n+1}(t) - x_{n-1}(t))$ $(n \geq 1)$, where we use $'$ for the derivative with respect to t, is known as the *Langmuir lattice* or the *Kac–van Moerbeke equation*. Our weight $w(x) = e^{-x^4+tx^2}$ is of the form given by the theorem for every $t \in \mathbb{R}$, so the recurrence coefficients of the Freud weight satisfy (2.14).

2.5 Painlevé IV

We will combine the discrete Painlevé equation (2.4) and the Kac–van Moerbeke equation (2.14). First, we set $x_n = a_n^2$ and write $x_n' = \frac{d}{dt}x_n$, so that

$$n = 4x_n(x_{n+1} + x_n + x_{n-1} - t/2) \tag{2.16}$$

$$x_n' = x_n(x_{n+1} - x_{n-1}). \tag{2.17}$$

Differentiate (2.17) to find

$$x_n'' = x_n'(x_{n+1} - x_{n-1}) + x_n(x_{n+1}' - x_{n-1}').$$

Replace x_{n+1}' and x_{n-1}' by using (2.17), then

$$x_n'' = x_n'(x_{n+1} - x_{n-1}) + x_n\Big(x_{n+1}(x_{n+2} - x_n) - x_{n-1}(x_n - x_{n-2})\Big).$$

The terms $x_{n+1}x_{n+2}$ and $x_{n-1}x_{n-2}$ can be replaced by using (2.16):

$$4x_{n+1}x_{n+2} = n + 1 - 4x_{n+1}(x_{n+1} + x_n - t/2),$$
$$4x_{n-1}x_{n-2} = n - 1 - 4x_{n-1}(x_n + x_{n-1} - t/2).$$

This gives

$$x_n'' = x_n'(x_{n+1} - x_{n-1}) - x_n^2(x_{n+1} + x_{n-1})$$
$$+ \frac{1}{2}x_n\Big(n - 2(x_{n+1}^2 + x_{n-1}^2) + (t - 2x_n)(x_{n+1} + x_{n-1})\Big).$$

Now eliminate $x_{n+1} + x_{n-1}$ and $x_{n+1} - x_{n-1}$ by using (2.16)–(2.17):

$$x_{n+1} + x_{n-1} = \frac{n}{4x_n} - x_n + \frac{t}{2},$$

$$x_{n+1} - x_{n-1} = \frac{x_n'}{x_n},$$

$$2(x_{n+1}^2 + x_{n-1}^2) = (x_{n+1} + x_{n-1})^2 + (x_{n+1} - x_{n-1})^2,$$

then one finds

$$x_n'' = \frac{(x_n')^2}{2x_n} + \frac{3x_n^3}{2} - x_n^2 t + x_n\left(\frac{n}{4} + \frac{t^2}{8}\right) - \frac{n^2}{32x_n}. \tag{2.18}$$

Exercise 4: Use the transformation $2x_n(t) = y(-t/2)$ and show that y satisfies the Painlevé IV equation in (1.21).

2.6 Orthogonal polynomials on a cross

So far we investigated the orthogonal polynomials with the weight function $e^{-x^4 + tx^2}$ on the real line, but the weight function is also positive and tends to zero fast on the imaginary axis $i\mathbb{R}$. So one could consider monic orthogonal polynomials $(Q_n)_n$ on the imaginary axis with this weight:

$$\int_{-i\infty}^{+i\infty} Q_n(x)Q_m(x)e^{-x^4 + tx^2}\, dx = 0, \qquad m \neq n.$$

It is not so hard to see that $Q_n(x) = (-i)^n P_n(ix)$, where $P_n(x) = p_n(x)/\gamma_n$ are the monic orthogonal polynomials on the real line with weight $e^{-x^4 - tx^2}$. So one changes the variable x to ix and t changes to $-t$. The recurrence relation (1.6)

$$P_{n+1}(x) = xP_n(x) - a_n^2(t)P_{n-1}(x),$$

readily gives for the $(Q_n)_n$

$$Q_{n+1}(x) = xQ_n(x) + a_n^2(-t)Q_{n-1}(x),$$

and we see that the recurrence coefficients for the $(Q_n)_n$ are given by $-a_n^2(-t)$. These recurrence coefficients are now negative, reflecting the fact that we are dealing with orthogonal polynomials on the imaginary axis. They correspond to the unique *negative* solution of d-P_I.

We can combine the real axis $\mathbb{R}$ and the imaginary axis $i\mathbb{R}$ and consider orthogonal polynomials on the cross $\mathbb{R} \cup i\mathbb{R}$. We can give a weight $\alpha > 0$ to the real axis and a weight $\beta > 0$ to the imaginary axis, and look for monic polynomials $(R_n)_n$ satisfying the orthogonality relations

$$\alpha \int_{-\infty}^{\infty} R_n(x)R_m(x)e^{-x^4 + tx^2}\, dx + \beta \int_{-i\infty}^{+i\infty} R_n(x)R_m(x)e^{-x^4 + tx^2}\, |dx| = 0, \qquad m \neq n.$$

Observe that we are not using complex conjugation in the bilinear form

$$\langle p, q \rangle = \alpha \int_{-\infty}^{\infty} p(x)q(x)w(x)\, dx + \beta \int_{-i\infty}^{+i\infty} p(x)q(x)w(x)\, |dx|$$

and since we are not on the real line anymore, it may happen that such polynomials R_n do not exist for every n. If R_n, R_{n-1} and R_{n+1} exist, then they will still be connected through a three term recurrence relation

$$R_{n+1}(x) = xR_n(x) - c_n R_{n-1}(x),$$

with recurrence coefficients $(c_n)_n$, and the proof of Theorem 2.1 still works, so that c_n satisfies d-P_I

$$4c_n(c_{n+1} + c_n + c_{n-1} - \frac{t}{2}) = n,$$

with $c_0 = 0$ but with a different initial condition for $n = 1$

$$c_1 = \frac{\alpha m_2(t) - \beta m_2(-t)}{\alpha m_0(t) + \beta m_0(-t)},$$

where

$$m_0(t) = \int_{-\infty}^{\infty} e^{-x^4 + tx^2}\, dx, \quad m_2(t) = \int_{-\infty}^{\infty} x^2 e^{-x^4 + tx^2}\, dx.$$

These first two moments can be expressed in terms of parabolic cylinder functions

$$m_0(t) = 2^{-1/4}\sqrt{\pi}\, e^{t^2/8}\, U(0, -t/\sqrt{2}), \qquad m_2(t) = \frac{d}{dt} m_0(t),$$

where the parabolic cylinder function is given by [123, §12.5]

$$U(a, z) = \frac{e^{-z^2/4}}{\Gamma(\frac{1}{2} + a)} \int_0^{\infty} t^{a-\frac{1}{2}} e^{-\frac{t^2}{2} - zt}\, dt.$$

The recurrence coefficients $(c_n)_n$ have a different behavior than those for the real line ($\beta = 0$) or the imaginary axis ($\alpha = 0$) only. As an example, we plotted the recurrence coefficients $(c_n)_n$ for $t = 0$ and $\beta/\alpha = 1/2$ in Figure 2.2. The recurrence coefficients can be computed starting from the initial values, and the resulting computations do not seem to be sensitive to the initial value for c_1. This sensitivity is only present when $\alpha = 0$ or $\beta = 0$.

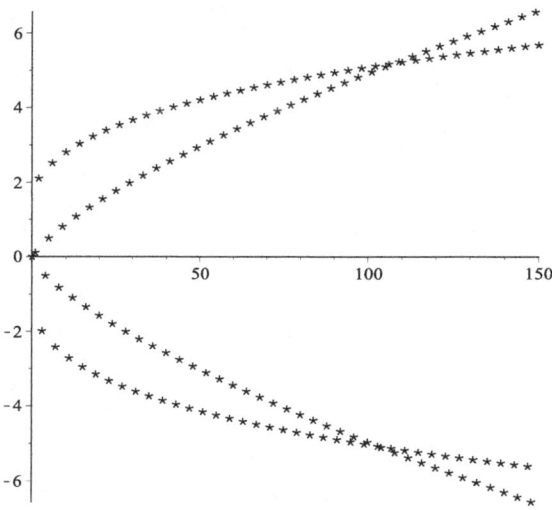

Figure 2.2 The recurrence coefficients c_n for $t = 0$ and $\beta/\alpha = 1/2$.

The case $\alpha = \beta$ and $t = 0$ is somewhat special. The initial condition becomes $c_1 = 0$, but this is a singularity for the discrete Painlevé equation and gives infinity for c_2. The singularity confinement however limits this singularity to a finite number of terms. What happens is that the orthogonal polynomials R_{4n-1} do not exist but the other polynomials exist and they are polynomials in x^4 (up to possible zeros at the origin). The zeros of R_n are all on the cross in a symmetrical way.

3

Discrete Painlevé II

3.1 Orthogonal polynomials on the unit circle

In this section we will consider orthogonal polynomials on the unit circle $\mathbb{T} = \{z \in \mathbb{C} : |z| = 1\}$. A very good source for the general theory is the recent set of books by Barry Simon [143]. The sequence of polynomials $\{\varphi_n, n = 0, 1, 2, \ldots\}$ is orthonormal on the unit circle with respect to a weight w if

$$\frac{1}{2\pi} \int_0^{2\pi} \varphi_n(z)\overline{\varphi_m(z)}w(\theta)\,d\theta = \delta_{n,m}. \tag{3.1}$$

These polynomials are unique if we agree to make the leading coefficient positive:

$$\varphi_n(z) = \kappa_n z^n + \cdots, \qquad \kappa_n > 0.$$

The monic polynomials are usually denoted by $\Phi_n(z) = \varphi_n(z)/\kappa_n$. The coefficient κ_n^2 can be expressed in terms of the trigonometric moments $(c_n)_{n\in\mathbb{Z}}$

$$c_n = \frac{1}{2\pi} \int_0^{2\pi} e^{-in\theta} w(\theta)\,d\theta,$$

as

$$\kappa_n^2 = \frac{D_{n-1}}{D_n},$$

where $D_n = \det(c_{i,i})_{i,j=0,\ldots,n}$ is the Toeplitz determinant for the weight w. An important property, which replaces the three term recurrence relation for orthogonal polynomials on the real line, is the Szegő recurrence

$$z\Phi_n(z) = \Phi_{n+1}(z) + \overline{\alpha_n}\Phi_n^*(z), \tag{3.2}$$

where $\Phi_n^*(z) = z^n\overline{\Phi}_n(1/z)$ is the reversed polynomial and $\overline{\Phi}_n$ is the polynomial Φ_n but with complex conjugated coefficients. In [143] the recurrence coefficients α_n ($n = 0, 1, 2, 3, \ldots$) are called *Verblunsky coefficients*. They are given by $\alpha_n = -\overline{\Phi_{n+1}(0)}$ and they satisfy $|\alpha_n| < 1$ for $n \geq 0$ and $\alpha_{-1} = -1$. Often one also uses the terminology *reflection coefficients* for $r_n = \Phi_n(0)$. An important relation between κ_n and α_n was found by Szegő [146, Thm. 11.3.2]:

$$\kappa_n^2 = \sum_{k=0}^{n} |\varphi_k(0)|^2,$$

from which it follows that

$$\kappa_{n+1}^2 - \kappa_n^2 = |\varphi_{n+1}(0)|^2 = \kappa_{n+1}^2 |\alpha_n|^2,$$

so that

$$\frac{\kappa_n^2}{\kappa_{n+1}^2} = 1 - |\alpha_n|^2, \tag{3.3}$$

and

$$|\alpha_n|^2 = \frac{D_n^2 - D_{n-1}D_{n+1}}{D_n^2}. \tag{3.4}$$

3.1.1 The weight $w(\theta) = e^{t \cos \theta}$

We will take a closer look at the orthogonal polynomials on the unit circle for the weight

$$w(\theta) = \exp(t \cos \theta), \qquad \theta \in [-\pi, \pi].$$

Observe that $w(-\theta) = w(\theta)$, which implies that the Verblunsky coefficients (reflection coefficients) are real. Ismail [91, pp. 236–239] calls the resulting orthogonal polynomials the *modified Bessel polynomials* since the trigonometric moments of this weight are in terms of the modified Bessel function

$$\frac{1}{2\pi} \int_0^{2\pi} e^{in\theta} w(\theta) \, d\theta = \frac{1}{\pi} \int_0^{\pi} \cos n\theta \exp(t \cos \theta) \, d\theta = I_n(t).$$

This weight and the corresponding orthogonal polynomials on the unit circle were investigated by Periwal and Shevitz [138] who were interested in unitary random matrices. They found a nonlinear recurrence relation for the Verblunsky coefficients of these orthogonal polynomials. If $z = e^{i\theta}$ then

$$w(\theta) = \exp\left(\frac{t}{2}\left(z + \frac{1}{z}\right)\right) := \hat{w}(z),$$

and this weight satisfies a Pearson equation of the form

$$\hat{w}'(z) = \frac{t}{2}(1 - \frac{1}{z^2})\hat{w}(z). \tag{3.5}$$

Note that the right-hand side contains a polynomial of degree 2 in $1/z$, rather than in z, but this is in fact good since the orthogonality is on the unit circle, where $\bar{z} = 1/z$. There is again a structure relation.

Property 3.1 *The monic orthogonal polynomials* $(\Phi_n)_{n\in\mathbb{N}}$ *on the unit circle for the weight* $w(\theta) = e^{t\cos\theta}$ *satisfy*

$$\Phi_n'(z) = n\Phi_{n-1}(z) + B_n\Phi_{n-2}(z), \tag{3.6}$$

with

$$B_n = \frac{t}{2}\frac{\kappa_{n-2}^2}{\kappa_n^2}. \tag{3.7}$$

Proof Expand Φ_n' in terms of the monic orthogonal polynomials, then

$$\Phi_n'(x) = \sum_{k=0}^{n-1} B_{n,k}\Phi_k(z).$$

Comparing coefficients of z^{n-1} gives $B_{n,n-1} = n$. For the other coefficients we multiply the expansion by $\overline{\Phi_k(z)}$ and integrate, to find

$$\kappa_k^{-2}B_{n,k} = \frac{1}{2\pi}\int_0^{2\pi} \Phi_n'(z)\overline{\Phi_k(z)}w(\theta)\,d\theta$$

$$= \frac{1}{2\pi i}\int_{\mathbb{T}} \Phi_n'(z)\overline{z\Phi_k(z)}\hat{w}(z)\,dz,$$

where we used the change of variable $z = e^{i\theta}$. Integration by parts (there are no boundary terms since we integrate over the closed contour $\mathbb{T}$) gives

$$\kappa_k^{-2}B_{n,k} = \frac{-1}{2\pi i}\int_{\mathbb{T}} \Phi_n(z)\left(\overline{z\Phi_k(z)}\hat{w}(z)\right)'\,dz.$$

A simple computation gives $(\overline{z\Phi_k(z)})' = -z^2[\overline{z\Phi_k(z)}]'$ for $z\in\mathbb{T}$, hence

$$\kappa_k^{-2}B_{n,k} = \frac{1}{2\pi i}\int_{\mathbb{T}} \Phi_n(z)\overline{z^2[z\Phi_k(z)]'}\hat{w}(z)\,dz - \frac{1}{2\pi i}\int_{\mathbb{T}} \Phi_n(z)\overline{z\Phi_k(z)}\hat{w}'(z)\,dz.$$

The first integral on the right is

$$\frac{1}{2\pi}\int_0^{2\pi} \Phi_n(z)\overline{z[z\Phi_k(z)]'}w(\theta)\,d\theta$$

and is zero by orthogonality whenever $k < n-1$. For the second integral on the right we use Pearson's equation (3.5) to find

$$\frac{1}{2\pi i}\int_{\mathbb{T}} \Phi_n(z)\overline{z\Phi_k(z)}\hat{w}'(z)\,dz = \frac{t}{4\pi i}\int_{\mathbb{T}} \Phi_n(z)\overline{z\Phi_k(z)}(1-z^2)\hat{w}(z)\,dz$$

$$= \frac{t}{4\pi}\int_0^{2\pi} \Phi_n(z)\overline{\Phi_k(z)}(1-z^2)w(\theta)\,d\theta$$

and this vanishes by orthogonality when $k < n - 2$. Hence only $B_{n,n-1} = n$ and $B_{n,n-2}$ are left in the Fourier expansion. We write $B_{n,n-2} = B_n$ and our previous calculations give

$$\kappa_{n-2}^{-2} B_n = -\frac{t}{4\pi} \int_0^{2\pi} \Phi_n(z)\overline{\Phi_{n-2}(z)(1 - z^2)}w(\theta)\,d\theta.$$

Clearly $\Phi_{n-2}(z)(1 - z^2) = -\Phi_n(z) +$ lower terms, hence

$$\kappa_{n-2}^{-2} B_n = \frac{t}{4\pi} \int_0^{2\pi} |\Phi_n(z)|^2 w(\theta)\,d\theta = \frac{t}{2}\frac{1}{\kappa_n^2},$$

which gives the required expression for B_n. □

The compatibility between the Szegő recurrence and the structure relation leads to the following result.

Theorem 3.2 (Periwal–Shevitz)   *The reflection coefficients (or the Verblunsky coefficients) $(\alpha_n)_{n\geq 0}$ for the orthogonal polynomials on the unit circle with weight function $w(\theta) = e^{t\cos\theta}$ satisfy the nonlinear equation*

$$-\frac{t}{2}(1 - \alpha_n^2)(\alpha_{n+1} + \alpha_{n-1}) = (n + 1)\alpha_n. \tag{3.8}$$

This nonlinear recurrence relation corresponds to the discrete Painlevé equation d-P_{II} in (1.25) with $\alpha_n = x_n$, $\alpha = \beta = -2/t$ and $a = 0$. The initial values are

$$\alpha_{-1} = -1, \qquad \alpha_0 = \frac{I_1(t)}{I_0(t)},$$

where I_1 and I_0 are modified Bessel functions.

Proof   Taking the derivative in the Szegő recurrence (3.2) gives

$$z\Phi_n'(z) + \Phi_n(z) = \Phi_{n+1}'(z) + \alpha_n(\Phi_n^*)'(z).$$

The reflection coefficients are all real, so that there is no need to use complex conjugation. Replace the derivatives Φ_n' and Φ_{n+1}' using the structure relation (3.6), then

$$z\Big(n\Phi_{n-1}(z) + B_n\Phi_{n-2}(z)\Big) + \Phi_n(z) = (n + 1)\Phi_n(z) + B_{n+1}\Phi_{n-1}(z) + \alpha_n(\Phi_n^*)'(z).$$

Now use the Szegő recurrence to replace $z\Phi_{n-1}(z)$ and $z\Phi_{n-2}(z)$, then

$$n\Big(\Phi_n(z) + \alpha_{n-1}\Phi_{n-1}^*(z)\Big) + B_n\Big(\Phi_{n-1}(z) + \alpha_{n-2}\Phi_{n-2}^*(z)\Big) + \Phi_n(z)$$
$$= (n + 1)\Phi_n(z) + B_{n+1}\Phi_{n-1}(z) + \alpha_n(\Phi_n^*)'(z).$$

Observe that the polynomial Φ_n drops out. We will expand the remaining polynomials using the monic polynomials $(\Phi_k)_{0 \leq n-1}$ but we will focus our attention on the coefficient for Φ_{n-1} in the expansion. Use

$$\Phi_n^*(z) = \Phi_n(0)z^n + \cdots = -\alpha_{n-1}\Phi_n(z) + \cdots ,$$

then equating the coefficients of Φ_{n-1} in the polynomial identity gives

$$-n\alpha_{n-1}\alpha_{n-2} + B_n = B_{n+1} - n\alpha_n\alpha_{n-1},$$

giving

$$B_{n+1} - B_n = n\alpha_{n-1}(\alpha_n - \alpha_{n-2}).$$

Use the expression (3.7) for B_n from Property 3.1, then

$$\frac{t}{2}\left(\frac{\kappa_{n-1}^2}{\kappa_{n+1}^2} - \frac{\kappa_{n-2}^2}{\kappa_n^2}\right) = n\alpha_{n-1}(\alpha_n - \alpha_{n-2}).$$

Finally use (3.3) to arrive at the desired equation (but for $n - 1$ instead of n).

For the initial conditions we have that $\alpha_{-1} = -\Phi_0 = -1$ and $\alpha_0 = -\Phi_1(0)$, so that $\Phi_1(z) = z - \alpha_0$. The orthogonality of Φ_1 and Φ_0 gives

$$\alpha_0 = \frac{\int zw(\theta)\,d\theta}{\int w(\theta)\,d\theta},$$

which gives the ratio of the modified Bessel functions $I_1(t)/I_0(t)$. $\quad\square$

3.1.2 The Ablowitz–Ladik lattice

The reflection coefficients (Verblunsky parameters) of orthogonal polynomials on the unit circle satisfy a system of differential-difference equations known as the *Ablowitz–Ladik lattice*.

Theorem 3.3 *Let μ be a positive measure on the unit circle which is symmetric (so that the reflection coefficients $(\alpha_n)_{n \in \mathbb{N}}$ are all real). Let μ_t be the measure such that $d\mu_t(\theta) = e^{t\cos\theta}\,d\mu(\theta)$, with $t \in \mathbb{R}$ a real parameter. The reflection coefficients $(\alpha_n(t))_{n \in \mathbb{N}}$ of the orthogonal polynomials on the unit circle for the measure μ_t then satisfy*

$$2\alpha_n' = (1 - \alpha_n^2)(\alpha_{n+1} - \alpha_{n-1}), \qquad n \geq 0, \qquad (3.9)$$

where $\alpha_n' = \frac{d}{dt}\alpha_n$.

Proof The norm of the nth monic orthogonal polynomial is

$$\int_0^{2\pi} |\Phi_n(z)|^2 e^{t\cos\theta}\,d\mu(\theta) = \frac{1}{\kappa_n^2}.$$

Taking derivatives with respect to t gives

$$\int_0^{2\pi} \left(\frac{d}{dt}\Phi_n(z)\right)\overline{\Phi_n(z)}\,d\mu_t(\theta) + \int_0^{2\pi} \Phi_n(z)\frac{d}{dt}\overline{\Phi_n(z)}\,d\mu_t(\theta)$$
$$+ \int_0^{2\pi} |\Phi_n(z)|^2 \cos\theta d\mu_t(\theta) = -2\frac{\kappa_n'}{\kappa_n^3}.$$

The polynomial Φ_n is monic, so its leading coefficient does not depend on t and therefore $\frac{d}{dt}\Phi_n$ is of degree $< n$. Hence orthogonality gives

$$-2\frac{\kappa_n'}{\kappa_n^3} = \int_0^{2\pi} |\Phi_n(z)|^2 \cos\theta\,d\mu_t(\theta) = \Re \int_0^{2\pi} z|\Phi_n(z)|^2\,d\mu_t(\theta).$$

Now use the recurrence relation (3.2) to find

$$\int_0^{2\pi} z|\Phi_n(z)|^2\,d\mu_t(\theta) = \int_0^{2\pi} \Phi_{n+1}(z)\overline{\Phi_n(z)}\,d\mu_t(\theta) + \alpha_n \int_0^{2\pi} \Phi_n^*(z)\overline{\Phi_n(z)}\,d\mu_t(\theta).$$

The first integral on the right vanishes by orthogonality. For the second integral we use $\Phi_n^*(z) = \Phi_n(0)z^n + \cdots + 1$ to find

$$\int_0^{2\pi} \Phi_n^*(z)\overline{\Phi_n(z)}\,d\mu_t(\theta) = \Phi_n(0) \int_0^{2\pi} |\Phi_n(z)|^2\,d\mu_t(\theta) = -\frac{\alpha_{n-1}}{\kappa_n^2}.$$

We thus find

$$2\frac{\kappa_n'}{\kappa_n} = \alpha_n\alpha_{n-1}. \qquad (3.10)$$

Next, we take derivatives with respect to t in (3.3) to find

$$\frac{\kappa_n^2}{\kappa_{n+1}^2}\left(\frac{\kappa_n'}{\kappa_n} - \frac{\kappa_{n+1}'}{\kappa_{n+1}}\right) = -\alpha_n\alpha_n'.$$

Use this in (3.10), then together with (3.3) we indeed find (3.9). $\qquad\square$

3.1.3 Painlevé V and III

We can now get a differential equation for α_n by combining (3.8) and (3.9)

$$\alpha_{n+1} + \alpha_{n-1} = \frac{-2(n+1)\alpha_n}{t(1-\alpha_n^2)}, \qquad (3.11)$$

$$\alpha_{n+1} - \alpha_{n-1} = \frac{2\alpha_n'}{1-\alpha_n^2}. \qquad (3.12)$$

Take derivatives in (3.9) to find

$$2\alpha_n'' = -2\alpha_n\alpha_n'(\alpha_{n+1} - \alpha_{n-1}) + (1-\alpha_n^2)(\alpha_{n+1}' - \alpha_{n-1}').$$

Replace α'_{n+1} and α'_{n-1} by using (3.9), then

$$2\alpha''_n = -2\alpha_n\alpha'_n(\alpha_{n+1}-\alpha_{n-1})+\frac{1}{2}(1-\alpha_n^2)\big((1-\alpha_{n+1}^2)(\alpha_{n+2}-\alpha_n)-(1-\alpha_{n-1}^2)(\alpha_n-\alpha_{n-2})\big).$$

Use (3.11) to replace

$$(1 - \alpha_{n+1}^2)\alpha_{n+2} = -(1 - \alpha_{n+1}^2)\alpha_n - \frac{2(n+2)}{t}\alpha_{n+1},$$

$$(1 - \alpha_{n-1}^2)\alpha_{n-2} = -(1 - \alpha_{n-1}^2)\alpha_n - \frac{2n}{t}\alpha_{n-1},$$

then

$$2\alpha''_n = -2\alpha_n\alpha'_n(\alpha_{n+1} - \alpha_{n-1}) + (1 - \alpha_n^2)\alpha_n(\alpha_{n+1}^2 + \alpha_{n-1}^2 - 2)$$
$$- \frac{(1 - \alpha_n^2)}{t}\big((n+1)(\alpha_{n+1} + \alpha_{n-1}) + \alpha_{n+1} - \alpha_{n-1}\big).$$

Now use (3.11)–(3.12) to eliminate α_{n+1} and α_{n-1}, then we have

$$\alpha''_n = -\frac{\alpha_n}{1 - \alpha_n^2}(\alpha'_n)^2 - \frac{\alpha'_n}{t} - \alpha_n(1 - \alpha_n^2) + \frac{(n+1)^2}{t^2}\frac{\alpha_n}{1 - \alpha_n^2}. \qquad (3.13)$$

Exercise 5: Use the rational transformation

$$\alpha_n = \frac{1+y}{1-y}$$

and show that y satisfies the Painlevé equation V

$$y'' = \left(\frac{1}{2y} + \frac{1}{y-1}\right)(y')^2 - \frac{y'}{t} + \frac{(n+1)^2}{8t^2}(y-1)^2\left(y - \frac{1}{y}\right) - \frac{2y(y+1)}{y-1}. \qquad (3.14)$$

Equation (3.14) is Painlevé V from (1.22) but with $\gamma = 0$. Such a P_V can be transformed to Painlevé III, see, e.g., [82, §41]. Tracy and Widom [147] and Hisakado [86] have obtained the corresponding P_{III} explicitly. Let

$$w_n = \frac{\alpha_n}{\alpha_{n-1}},$$

then

$$w'_n = \frac{\alpha'_n\alpha_{n-1} - \alpha'_{n-1}\alpha_n}{\alpha_{n-1}^2}$$

$$= \frac{\alpha'_n}{\alpha_{n-1}} - w_n\frac{\alpha'_{n-1}}{\alpha_{n-1}}. \qquad (3.15)$$

Subtracting (3.11) and (3.12) gives

$$\alpha_n' = -\frac{n+1}{t}\alpha_n - (1 - \alpha_n^2)\alpha_{n-1}, \tag{3.16}$$

and adding (3.11) and (3.12) gives (after changing n to $n-1$)

$$\alpha_{n-1}' = \frac{n}{t}\alpha_{n-1} + (1 - \alpha_{n-1}^2)\alpha_n. \tag{3.17}$$

Using (3.16)–(3.17) in (3.15) then gives

$$w_n' = -\frac{2n+1}{t}w_n - w_n^2 + 2\alpha_n^2 - 1. \tag{3.18}$$

Next, use (3.16) to find

$$(\alpha_n^2)' = 2\alpha_n\alpha_n' = -\frac{2(n+1)}{t}\alpha_n^2 - \frac{2\alpha_n^2(1 - \alpha_n^2)}{w_n},$$

and then, taking the derivative of (3.18) gives

$$w_n'' = -\frac{2n+1}{t}w_n' + \frac{2n+1}{t^2}w_n - \frac{4(n+1)}{t}\alpha_n^2 - \frac{4\alpha_n^2(1 - \alpha_n^2)}{w_n} - 2w_n w_n'.$$

From (3.18) we can find α_n^2 in terms of w_n' and w_n, and then using this, one obtains after some simple algebra

$$w_n'' = \frac{(w_n')^2}{w_n} - \frac{w_n'}{t} + \frac{2n}{t}w_n^2 - \frac{2(n+1)}{t} + w_n^3 - \frac{1}{w_n}. \tag{3.19}$$

This is the Painlevé III equation in (1.20) with $\alpha = 2n$, $\beta = -2(n+1)$, $\gamma = 1$ and $\delta = -1$.

3.2 Discrete orthogonal polynomials

In this section we will study certain discrete orthogonal polynomials on the integers $\mathbb{N}$. The orthonormality now becomes

$$\sum_{k=0}^{\infty} p_n(k)p_m(k)w_k = \delta_{n,m}, \qquad n, m \geq 0. \tag{3.20}$$

Instead of the differential operator we will now be using difference operators, namely the forward difference Δ and the backward difference ∇ for which

$$\Delta f(x) = f(x+1) - f(x), \qquad \nabla f(x) = f(x) - f(x-1).$$

We now have two sequences $(a_n)_{n=1,2,\ldots}$ and $(b_n)_{n=0,1,2,\ldots}$ of recurrence coefficients, and we need two recurrence relations to determine all a_n and b_n.

Charlier polynomials are the orthonormal polynomials for the Poisson distribution

$$w_k = \frac{a^k}{k!}, \qquad k \in \mathbb{N}, \ a > 0.$$

Observe that

$$w_{k-1} = \frac{k}{a} w_k, \tag{3.21}$$

which is the (discrete) Pearson equation for the Poisson distribution. It can also be written as $a \nabla w_k = (a - k) w_k$, which is a special form of the general Pearson equation $\nabla(\sigma w) = \tau w$ for the difference operator, with σ and τ polynomials. The Pearson equation gives the following structure relation for Charlier polynomials.

Property 3.4 *For the orthonormal Charlier polynomials one has*

$$p_n(x + 1) = p_n(x) + A_n p_{n-1}(x), \tag{3.22}$$

or $\Delta p_n(x) = A_n p_{n-1}(x)$.

Proof If we expand $p_n(x + 1)$ into a Fourier series, then

$$p_n(x + 1) = \sum_{j=0}^{n} A_{n,j} p_j(x),$$

and if we compare the leading coefficients then $A_{n,n} = 1$. The other Fourier coefficients are given by

$$A_{n,j} = \sum_{k=0}^{\infty} p_n(k + 1) p_j(k) w_k = \sum_{k=1}^{\infty} p_n(k) p_j(k - 1) w_{k-1},$$

and if we use (3.21) then this gives

$$A_{n,j} = \frac{1}{a} \sum_{k=0}^{\infty} k p_n(k) p_j(k - 1) w_k.$$

The polynomial $x p_j(x - 1)$ has degree $j + 1$, hence by orthogonality $A_{n,j} = 0$ whenever $j < n - 1$. $\qquad \square$

Again the three term recurrence relation and the structure relation need to be compatible. Apply the forward difference operator Δ to (1.2), then

$$(x + 1) p_n(x + 1) - x p_n(x) = a_{n+1} \Delta p_{n+1}(x) + b_n \Delta p_n(x) + a_{n-1} \Delta p_{n-1}(x).$$

Then use the structure relation (3.22) to find

$$(x + 1) A_n p_{n-1}(x) + p_n(x) = a_{n+1} A_{n+1} p_n(x) + b_n A_n p_{n-1}(x) + a_n A_{n-1} p_{n-2}(x).$$

Use the recurrence relation (1.2) for $xp_{n-1}(x)$, then

$$A_n p_{n-1}(x) + A_n \big(a_n p_{n-1}(x) + b_{n-1} p_{n-1}(x) + a_{n-1} p_{n-2}(x) \big) + p_n(x)$$
$$= a_{n+1} A_{n+1} p_n(x) + b_n A_n p_{n-1}(x) + a_n A_{n-1} p_{n-2}(x).$$

Since p_n, p_{n-1} and p_{n-2} are linearly independent polynomials, this can only be true if the coefficients before these three polynomials are zero, giving

$$p_n(x) \Rightarrow A_n a_n + 1 = a_{n+1} A_{n+1}, \qquad (3.23)$$

$$p_{n-1}(x) \Rightarrow A_n + A_n b_{n-1} = b_n A_n, \qquad (3.24)$$

$$p_{n-2}(x) \Rightarrow A_n a_{n-1} = a_n A_{n-1}. \qquad (3.25)$$

From (3.23) we find $A_n a_n = n$ and from (3.25) we see that $A_n / a_n = A_{n-1} / a_{n-1}$, hence this ratio is constant and $A_n = c a_n$, with c a constant. We thus have $a_n^2 n / c$ and for $n = 1$ we conclude that $a_n^2 = n a_1^2$. From (3.24) we find $b_n - b_{n-1} = 1$ so that $b_n = n + b_0$. The only thing left is to determine b_0 and a_1^2. Introduce the moments

$$m_n = \sum_{k=0}^{\infty} k^n \frac{a^k}{k!},$$

then $p_0^2 = 1/m_0$, $b_0 = m_1/m_0$ and $a_1^2 = m_2/m_0 - (m_1/m_0)^2$. The first few moments are

$$m_0 = e^{-a}, \qquad m_1 = a e^{-a}, \qquad m_2 = (a^2 + a) e^{-a}$$

so that $b_0 = a$ and $a_1^2 = a$. These simple computations show that the recurrence coefficients for Charlier polynomials are given by

$$a_n = \sqrt{an}, \qquad b_n = n + a.$$

3.2.1 Generalized Charlier polynomials

If we take the weight

$$w_k = \frac{a^k}{k!(\beta)_k}, \qquad k \in \mathbb{N}, \ a > 0, \beta > 0, \qquad (3.26)$$

then we get generalized Charlier polynomials. Here we used the Pochhammer symbol (rising factorial)

$$(\beta)_k = \beta(\beta + 1)(\beta + 2) \cdots (\beta + k - 1).$$

The case $\beta = 1$, for which $w_k = a^k/(k!)^2$, was introduced by Hounkonnou et al. in [87] and the nonlinear recurrence relation was found in [76]. The case $\beta \neq 1$

was investigated in [144]. Observe that

$$w_{k-1} = \frac{k(\beta + k - 1)}{a} w_k,$$

which can also be written as $a\nabla w_k = (a - (\beta - 1)k - k^2)w_k$. The polynomial τ is of degree 2 and hence this is a semi-classical discrete weight. We can use the weight function

$$w(x) = \frac{a^x \Gamma(\beta)}{\Gamma(x+1)\Gamma(x+\beta)},$$

for which $w_k = w(k)$, and this satisfies the Pearson equation

$$a\nabla w(x) = (a - (\beta - 1)x - x^2)w(x), \tag{3.27}$$

with boundary condition $w(-1) = 0$.

Property 3.5 *Let w be a positive weight satisfying the Pearson equation*

$$\nabla[\sigma(x)w(x)] = \tau(x)w(x),$$

with initial condition $w(-1) = 0$, where σ and τ are polynomials. Then the orthogonal polynomials on $\mathbb{N}$ for this weight satisfy the structure relation

$$\sigma(x)\Delta p_n(x) = \sum_{j=n-t}^{n+s-1} A_{n,j} p_j(x), \tag{3.28}$$

where $s = \deg \sigma$ and $t = \max\{\deg \tau, \deg \sigma - 1\}$.

Exercise 6: Prove this structure relation for discrete orthogonal polynomials.

Hint: use summation by parts

$$\sum_{k=0}^{\infty} g(k)\Delta f(k) = -\sum_{k=0}^{\infty} f(k)\nabla g(k),$$

which holds whenever $g(-1) = 0$ and

$$\sum_{k=0}^{\infty} |f(k)| < \infty, \qquad \sum_{k=0}^{\infty} |g(k)| < \infty.$$

Property 3.6   *The recurrence coefficients for the orthogonal polynomials with weights (3.26) on* $\mathbb{N}$ *satisfy*

$$b_n + b_{n-1} - n + \beta = \frac{an}{a_n^2}, \tag{3.29}$$

$$b_{n-1} - b_n + 1 = \frac{a_n^2}{an}(a_{n+1}^2 - a_{n-1}^2), \tag{3.30}$$

with initial conditions $a_0^2 = 0$ *and*

$$b_0 = \frac{\sqrt{a}I_\beta(2\sqrt{a})}{I_{\beta-1}(2\sqrt{a})},$$

where I_ν *is the modified Bessel function*

$$I_\nu(z) = \sum_{k=0}^{\infty} \frac{(z/2)^{2k+\nu}}{k!\Gamma(k+\nu+1)}.$$

Proof   The structure relation for the orthonormal polynomials for the generalized Charlier weight (3.26) is given by

$$\Delta p_n(x) = A_n p_{n-1}(x) + B_n p_{n-2}(x). \tag{3.31}$$

Apply the forward difference operator Δ to the three term recurrence relation (1.2) to find

$$(x+1)p_n(x+1) - xp_n(x) = a_{n+1}\Delta p_{n+1}(x) + b_n\Delta p_n(x) + a_n\Delta p_{n-1}(x).$$

The left-hand side can also be written as $(x+1)\Delta p_n(x) + p_n(x)$. Use the structure relation (3.31) to replace $\Delta p_n(x)$, $\Delta p_{n+}(x)$ and $\Delta p_{n-1}(x)$, then

$$(x+1)\Big(A_n p_{n-1}(x) + B_n p_{n-2}(x)\Big) + p_n(x) = a_{n+1}\Big(A_{n+1}p_n(x+1) + B_{n+1}p_{n-1}(x)\Big)$$
$$+ b_n\Big(A_n p_{n-1}(x) + B_n p_{n-2}(x)\Big) + a_n\Big(A_{n-1}p_{n-2}(x) + B_{n-1}p_{n-3}(x)\Big).$$

Now use the three term recurrence relation (1.2) to replace $xp_{n-1}(x)$ and $xp_{n-2}(x)$ to find

$$A_n\Big(a_n p_n(x) + b_{n-1}p_{n-1}(x) + a_{n-1}p_{n-2}(x)\Big)$$
$$+ B_n\Big(a_{n-1}p_{n-1}(x) + b_{n-2}p_{n-2}(x) + a_{n-2}p_{n-3}(x)\Big)$$
$$+ A_n p_{n-1}(x) + B_n p_{n-2}(x) + p_n(x) = a_{n+1}\Big(A_{n+1}p_n(x+1) + B_{n+1}p_{n-1}(x)\Big)$$
$$+ b_n\Big(A_n p_{n-1}(x) + B_n p_{n-2}(x)\Big) + a_n\Big(A_{n-1}p_{n-2}(x) + B_{n-1}p_{n-3}(x)\Big).$$

This is a linear combination of the four polynomials $p_n(x), p_{n-1}(x), p_{n-2}(x)$

and $p_{n-3}(x)$, hence this can only be true if the coefficients in front of these polynomials are zero, giving

$$p_n(x) \Rightarrow A_n a_n + 1 = a_{n+1} A_{n+1}, \tag{3.32}$$

$$p_{n-1}(x) \Rightarrow A_n b_{n-1} + B_n a_{n-1} + A_n = a_{n+1} B_{n+1} + b_n A_n, \tag{3.33}$$

$$p_{n-2}(x) \Rightarrow A_n a_{n-1} + B_n b_{n-2} + B_n = b_n B_n + a_n A_{n-1}, \tag{3.34}$$

$$p_{n-3}(x) \Rightarrow B_n a_{n-2} = a_n B_{n-1}. \tag{3.35}$$

As before (3.32) gives that $A_n a_n = n$. From (3.35) we get the relation $B_n/a_n a_{n-1} = B_{n-1}/a_{n-1} a_{n-2}$ so that this ratio is constant, giving $B_n = c a_n a_{n-1}$, with c a constant. From (3.33) we find

$$A_n(b_{n-1} - b_n + 1) = c a_n(a_{n+1}^2 - a_{n-1}^2)$$

and if we multiply this by a_n, then

$$n(b_{n-1} - b_n + 1) = c a_n^2(a_{n+1}^2 - a_{n-1}^2). \tag{3.36}$$

From (3.34) we find

$$c(b_n - b_{n-2} - 1) = \frac{A_n}{a_n} - \frac{A_{n-1}}{a_{n-1}},$$

and summing from 2 to n gives

$$\frac{A_n}{a_n} = \frac{A_1}{a_1} + c(b_n + b_{n-1} - b_0 - b_1 - n + 1).$$

Multiplying by a_n^2 then gives

$$n = \frac{a_n^2}{a_1^2} + c a_n^2(b_n + b_{n-1} - b_0 - b_1 - n + 1). \tag{3.37}$$

The equations (3.36)–(3.37) still contain constants c, b_0, b_1, a_1^2. If we multiply the structure relation (3.31) for $n = 1$ by $p_0 w(x)$ and sum over $x \in \mathbb{N}$ then

$$A_1 = p_0 \sum_{k=0}^{\infty} \Delta p_1(k) w_k = -p_0 \sum_{k=0}^{\infty} p_1(k) \nabla w_k,$$

where we used summation by parts. The Pearson equation (3.27) gives

$$A_1 = -\frac{p_0}{a} \sum_{k=0}^{\infty} p_1(k)(a - (\beta - 1)k - k^2) w_k$$

$$= \frac{1}{a}(\beta - 1) \sum_{k=0}^{\infty} k p_1(k) p_0 w_k + \frac{1}{a} \sum_{k=0}^{\infty} k^2 p_1(k) p_0 w_k.$$

The first sum is $(\beta - 1)a_1/a$ by using (1.4), for the second sum we use the three term recurrence relation and find $a_1(b_0 + b_1)/a$, hence

$$A_1 = \frac{a_1}{a}(\beta - 1 + b_0 + b_1),$$

and recall that $a_1 A_1 = 1$, hence $a_1^2(b_0 + b_1 + \beta - 1) = a$. In a similar way we can multiply the structure relation (3.31) for $n = 2$ by $p_0 w(x)$ and sum, to get

$$B_2 = p_0 \sum_{k=0}^{\infty} \Delta p_2(k)\, w_k = -p_0 \sum_{k=0}^{\infty} p_2(k) \nabla w_k$$

and the Pearson equation (3.27) gives

$$B_2 = -\frac{p_0}{a} \sum_{k=0}^{\infty} p_2(k)(a - (\beta - 1)k - k^2)w_k = \frac{1}{a} \sum_{k=0}^{\infty} k^2 p_2(k) p_0 w_k.$$

With a little help from the three term recurrence relation, this gives $B_2 = a_1 a_2/a$ so that the constant c is equal to $1/a$. If we use all this information in (3.36)–(3.37), then we arrive at the desired equations (3.29)–(3.30).   □

Observe that (3.32) is a first order equation for b_n but (3.33) is a second order equation for a_n. We have to do a little bit more work to arrive at two first order equations or one second order equation.

Theorem 3.7   *The recurrence coefficients of the generalized Charlier polynomials with weight (3.26) satisfy*

$$(a_{n+1}^2 - a)(a_n^2 - a) = a d_n(d_n + \beta - 1), \qquad (3.38)$$

$$d_n + d_{n-1} + n + \beta - 1 = \frac{an}{a_n^2}, \qquad (3.39)$$

where $d_n = b_n - n$.

Proof   From (3.32) we find $an = a_n^2(b_n + b_{n-1} - n + \beta)$. Insert this in (3.33) to find

$$(b_{n-1} - b_n + 1)(b_n + b_{n-1} - n + \beta) = a_{n+1}^2 - a_{n-1}^2.$$

We now introduce a new sequence $(d_n)_{n \geq 0}$ by putting $b_n = n + d_n$, then the last equation becomes

$$(d_{n-1} - d_n)(d_n + d_{n-1} + n + \beta - 1) = a_{n+1}^2 - a_{n-1}^2,$$

which is the same as

$$(d_{n-1}^2 - d_n^2) + n(d_{n-1} - d_n) + (\beta - 1)(d_{n-1} - d_n) = a_{n+1}^2 - a_{n-1}^2. \qquad (3.40)$$

On the other hand, (3.33) is

$$an(d_{n-1} - d_n) = a_n^2 a_{n+1}^2 - a_{n-1}^2 a_n^2,$$

so if we insert this in (3.40), then we have

$$(d_{n-1}^2 - d_n^2) + \frac{1}{a}(a_n^2 a_{n+1}^2 - a_{n-1}^2 a_n^2) + (\beta - 1)(d_{n-1} - d_n) = a_{n+1}^2 - a_{n-1}^2.$$

This equation only contains differences, so summing from 1 to n gives

$$d_0^2 - d_n^2 + \frac{1}{a}a_n^2 a_{n+1}^2 + (\beta - 1)(d_0 - d_n) = a_{n+1}^2 + a_n^2 - a_1^2.$$

This equation still has $d_0 = b_0$ and a_1^2. By Parseval's identity we have for the moments m_k

$$\frac{m_2}{m_0} = \sum_{k=0}^{\infty} [k p_0(k)]^2 w_k = a_1^2 + b_0^2,$$

and on the other hand, summing Pearson's equation (3.27) gives

$$0 = a m_0 - (\beta - 1) m_1 - m_2.$$

Recalling that $b_0 = m_1/m_0$, we thus have

$$a_1^2 + b_0^2 = \frac{m_2}{m_0} = a - (\beta - 1) b_0.$$

Hence we have

$$\frac{a_n^2 a_{n+1}^2}{a} - a_{n+1}^2 - a_n^2 + a = d_n^2 + (\beta - 1) d_n.$$

The left-hand side can be factored and gives (3.38). The other equation (3.39) is the same as (3.29) but with $b_n = d_n + n$.   □

The case $\beta = 1$ can be worked out in more detail. Indeed, when $\beta = 1$ we have

$$(a_{n+1}^2 - a)(a_n^2 - a) = a d_n^2, \tag{3.41}$$

$$d_n + d_{n-1} + n = \frac{an}{a_n^2}, \tag{3.42}$$

hence $a_n^2 - a$ has constant sign for all $n \geq 0$. Then $a_0 = 0$ implies that $a_n^2 - a < 0$ for all $n \geq 0$, hence we can introduce a sequence $(c_n)_{n \geq 0}$ with $c_0 = 1$ and $a_n^2 - a = -a c_n^2$. Insert this in (3.41), then $a c_n^2 c_{n+1}^2 = d_n^2$, hence we have $d_n = \sqrt{a} c_n c_{n+1}$, where we recursively choose the sign of c_{n+1} equal to the sign of $d_n c_n$. Insert this in (3.42), then

$$\sqrt{a} c_n (c_{n+1} + c_{n-1}) + n = \frac{n}{1 - c_n^2},$$

which readily gives

$$c_{n+1} + c_{n-1} = \frac{nc_n}{\sqrt{a}(1 - c_n^2)}. \tag{3.43}$$

This is the discrete Painlevé equation d-P_{II} in (1.25), with $\alpha = 1/\sqrt{a}, \beta = 0$.

The case $\beta = 1/2$ has a simple solution, namely

$$a_n^2 = \frac{n\sqrt{a}}{2}, \qquad b_n = \frac{n}{2} + \sqrt{a},$$

with initial values $a_0^2 = 0$ and $b_0 = \sqrt{a}$. This corresponds to the recurrence coefficients of Charlier polynomials on the half integers $\{0, 1/2, 1, 3/2, 2, \ldots\}$ with parameter $2\sqrt{a}$. Observe that our generalized Charlier weight on the half integers is

$$w(k/2) = \frac{a^{k/2}\sqrt{\pi}}{\Gamma(k/2 + 1)\Gamma((k + 1)/2)},$$

and Legendre's duplication formula for the Gamma function

$$\Gamma(z)\Gamma(z + 1/2) = \sqrt{\pi}2^{1-2z}\Gamma(2z)$$

then gives

$$w(k/2) = \frac{(2\sqrt{a})^k}{(2k)!},$$

which are the weights for the Poisson distribution with parameter $2\sqrt{a}$. Since we are now using orthogonality on $\frac{1}{2}\mathbb{N}$ one needs to divide the recurrence coefficients of the usual Charlier polynomials by 2.

For all other cases the recurrence relations (3.38)–(3.39) are a limiting case of the discrete Painlevé equation corresponding to D_4^c in Sakai's classification, or d-$P(D_4^{(1)}/D_4^{(1)})$ in the terminology of [96].

Exercise 7: One of the discrete Painlevé equations for D_4^c is given by the system

$$x_{n+1}x_n = \frac{(y_n - z_n)^2 - A}{y_n^2 - B},$$

$$y_n + y_{n-1} = \frac{z_{n-1/2} - C}{1 + Dx_n} + \frac{z_{n-1/2} + C}{1 + x_n/D}$$

with $z_n = \alpha n + \beta$. Show that for $x_n = iX_n/\sqrt{aB}$ and $iD = \sqrt{B/a}$ and $B \to \infty$ we arrive at a discrete Painlevé equation of the form (3.38)–(3.39). Identify the X_n, y_n and the parameters.

3.2.2 The Toda lattice

Theorem 3.8 *Suppose μ is a positive measure on the real line for which all the moments exist, and μ_t is the modified measure with $d\mu_t(x) = e^{xt} d\mu_t(x)$, where $t \in \mathbb{R}$ is such that all the moments of μ_t exist. Then the recurrence coefficients $(a_n)_{n \geq 1}$ and $(b_n)_{n \geq 0}$ of the orthogonal polynomials for the measure μ_t satisfy*

$$\frac{d}{dt}a_n^2 = a_n^2(b_n - b_{n-1}), \qquad n \geq 1, \tag{3.44}$$

$$\frac{d}{dt}b_n = a_{n+1}^2 - a_n^2, \qquad n \geq 0, \tag{3.45}$$

with $a_0^2 = 0$.

Proof   The three term recurrence relation (1.2) is

$$xp_n(x; t) = a_{n+1}(t)p_{n+1}(x; t) + b_n(t)p_n(x; t) + a_n(t)p_{n-1}(x; t),$$

where we denoted the dependence on the parameter t explicitly. Taking derivatives with respect to t gives

$$x\frac{d}{dt}p_n(x; t) = \left(\frac{d}{dt}a_{n+1}\right)p_{n+1}(x; t) + \left(\frac{d}{dt}b_n\right)p_n(x; t) + \left(\frac{d}{dt}a_n\right)p_{n-1}(x; t)$$

$$+ a_{n+1}\frac{d}{dt}p_{n+1}(x; t) + b_n\frac{d}{dt}p_n(x; t) + a_n\frac{d}{dt}p_{n-1}(x; t).$$

Multiply this by $p_{n+1}(x; t)$ and integrate with respect to the measure μ_t, then

$$\int x\left(\frac{d}{dt}p_n(x; t)\right)p_{n+1}(x; t)\, d\mu_t(x)$$

$$= \frac{d}{dt}a_{n+1} + a_{n+1}\int \left(\frac{d}{dt}p_{n+1}(x; t)\right)p_{n+1}(x; t)\, d\mu_t(x),$$

where we used the orthonormality of the polynomials. Now use the three term recurrence relation for $xp_{n+1}(x; t)$ on the left-hand side, then

$$\frac{d}{dt}a_{n+1} = a_{n+1}\left(\int p_n(x; t)\left(\frac{d}{dt}p_n(x; t)\right) d\mu_t(x)\right.$$

$$\left. - \int p_{n+1}(x; t)\left(\frac{d}{dt}p_{n+1}(x; t)\right) d\mu_t(x)\right). \tag{3.46}$$

Integrating the orthonormality relation

$$\int p_n^2(x; t)e^{xt}\, d\mu(x) = 1$$

gives

$$2 \int p_n(x;t) \left(\frac{d}{dt} p_n(x;t) \right) d\mu_t(x) + \int x p_n^2(x;t) \, d\mu_t(x) = 0,$$

which by (1.5) gives

$$2 \int p_n(x;t) \left(\frac{d}{dt} p_n(x;t) \right) d\mu_t(x) = -b_n.$$

Insert this in (3.46) to find (3.44).   □

Exercise 8: Complete the proof of Theorem 3.8 by giving a proof of (3.45).

The equations (3.44)–(3.45) are known as the *Toda lattice*. This is an integrable system with a nonlinear Hamiltonian, first introduced by Toda at the end of the 1960s.

3.2.3 Painlevé V and III

If we put $a = e^t$ in the weight (3.26), then we get a measure of the form $d\mu_t(x) = e^{tx} \, d\mu(x)$ where μ is the discrete measure on $\mathbb{N}$ with weights given by $1/[k!(\beta)_k]$, i.e., the generalized Charlier weights with $a = 1$. Hence the recurrence coefficients, as a function of $t = \log a$, satisfy the Toda equations (3.44)–(3.45). They also satisfy the nonlinear equations (3.38)–(3.39). We will study the recurrence coefficients a_n^2 and b_n as a function of a rather than as a function of t, and we will use $x_n(a) = a_n^2$ and $y_n(a) = b_n$ and $x_n' = dx_n/da$ and $y_n' = dy_n/da$. Note that

$$x_n' = \frac{1}{a} \frac{dx_n}{dt}, \qquad y_n' = \frac{1}{a} \frac{dy_n}{dt},$$

so the Toda equations become

$$x_n' = \frac{x_n}{a}(y_n - y_{n-1}), \tag{3.47}$$

$$y_n' = \frac{1}{a}(x_{n+1} - x_n), \tag{3.48}$$

and the nonlinear equations are

$$(x_n - a)(x_{n+1} - a) = a(y_n - n)(y_n - n + \beta - 1), \tag{3.49}$$

$$y_n + y_{n-1} - n + \beta = \frac{an}{x_n}. \tag{3.50}$$

If we take a derivative (with respect to a) of (3.47) then

$$x_n'' = \frac{a x_n' - x_n}{a^2}(y_n - y_{n-1}) + \frac{x_n}{a}(y_n' - y_{n-1}').$$

Use (3.47) to replace $y_n - y_{n-1}$ and (3.48) to replace y_n' and y_{n-1}' to find

$$x_n'' = \left(x_n' - \frac{x_n}{a}\right)\frac{x_n'}{x_n} + \frac{x_n}{a^2}(x_{n+1} - 2x_n + x_{n-1})$$

$$= \left(x_n' - \frac{x_n}{a}\right)\frac{x_n'}{x_n} - \frac{2x_n(x_n - a)}{a^2} + \frac{x_n}{a^2}(x_{n+1} + x_{n-1} - 2a). \qquad (3.51)$$

Add (3.49) for n and $n - 1$ to find

$$(x_n - a)(x_{n+1} + x_{n-1} - 2a) = a(y_n - n)(y_n - n + \beta - 1) + a(y_{n-1} - n + 1)(y_{n-1} - n + \beta).$$

The right-hand side can be written as

$$(y_n - n)(y_n - n + \beta - 1) + (y_{n-1} - n + 1)(y_{n-1} - n + \beta)$$
$$= \frac{1}{2}(y_n + y_{n-1} - 2n)^2 + \frac{1}{2}(y_n - y_{n-1})^2 + \beta(y_n + y_{n-1} - 2n + 1) - (y_n - y_{n-1}).$$

Use (3.50) to replace $y_n + y_{n-1}$ and (3.47) to replace $y_n - y_{n-1}$, then

$$(y_n - n)(y_n - n + \beta - 1) + (y_{n-1} - n + 1)(y_{n-1} - n + \beta)$$
$$= \frac{1}{2}\left(\frac{an}{x_n} - n - \beta\right)^2 + \frac{1}{2}\left(a\frac{x_n'}{x_n}\right)^2 + \beta\left(\frac{an}{x_n} - n - \beta + 1\right) - a\frac{x_n'}{x_n}.$$

If we insert this in (3.51), then

$$x_n'' = \left(x_n' - \frac{x_n}{a}\right)\frac{x_n'}{x_n} - \frac{2x_n(x_n - a)}{a^2}$$
$$+ \frac{x_n}{a(x_n - a)}\left(\frac{1}{2}\left(\frac{an}{x_n} - n - \beta\right)^2 + \frac{1}{2}\left(a\frac{x_n'}{x_n}\right)^2 + \beta\left(\frac{an}{x_n} - n - \beta + 1\right) - a\frac{x_n'}{x_n}\right).$$

This is a second order nonlinear differential equation for x_n, which we wish to identify as a Painlevé equation. Collecting terms gives

$$x_n'' = \frac{1}{2}\left(\frac{1}{x_n} + \frac{1}{x_n - a}\right)(x_n')^2 - \frac{x_n}{a(x_n - a)}x_n'$$
$$- \frac{2x_n(x_n - a)}{a^2} + \frac{an^2}{2}\frac{1}{x_n(x_n - a)} - \frac{n^2}{x_n - a} + \frac{n^2 - \beta^2 + 2\beta}{2a}\frac{x_n}{x_n - a}.$$

This equation contains $(x_n')^2$ and has singularities when $x_n = 0$ and $x_n = a$, hence it looks like Painlevé V in (1.22). However P_V has singularities as $y = 0$ and $y = 1$. If we introduce $x_n(a) = a/(1 - y(a))$, then

$$x_n' = \frac{1 - y + ay'}{(1 - y)^2}, \quad x_n'' = \frac{ay''}{(1 - y)^2} + \frac{2y'}{(1 - y)^2} + \frac{2a(y')^2}{(1 - y)^3},$$

and after some calculations we then find

$$y'' = \frac{1}{2}\left(\frac{1}{2y} + \frac{1}{y-1}\right)(y')^2 - \frac{y'}{a} + \frac{(1-y)^2}{a^2}\left(\frac{n^2 y}{2} - \frac{(\beta-1)^2}{2y}\right) - \frac{2y}{a},$$

and this is Painlevé V in (1.22) with $\alpha = n^2/2$, $\beta \to -(\beta-1)^2/2$, $\gamma = -2$ and $\delta = 0$. The Painlevé V equation with $\delta = 0$ can be reduced to Painlevé III. One has the following result (see, e.g., [82, Thm. 34.1 and 34.3]):

Theorem 3.9 *Suppose that $v(x)$ is a solution of P_{III} given in (1.20) with parameters $\alpha = a$, $\beta = b$, $\gamma = 1$ and $\delta = -1$, and*

$$y(x) = v'(x) \pm v^2(x) + \frac{(1 \pm a)}{x} v(x).$$

Then

$$w(z) = \frac{y(x) - 1}{y(x) + 1}, \qquad z = \frac{x^2}{2},$$

satisfies P_V given in (1.22), with parameters

$$\alpha = \frac{(b \pm a + 2)^2}{32}, \qquad \beta = -\frac{(b \mp a - 2)^2}{32}, \qquad \gamma = \pm 1, \qquad \delta = 0.$$

Conversely, if $w(z)$ is a solution of P_V in (1.22) with parameters α, β, $\gamma = \pm 1$ and $\delta = 0$ and

$$\phi(z) = zw' - \sqrt{2\alpha}w^2 + (\sqrt{2\alpha} + \sqrt{-2\beta})w - \sqrt{-2\beta},$$

then

$$v(x) = \frac{\sqrt{2z}w}{\phi(z)}, \qquad x^2 = 2z,$$

is a solution of $P_{III}(a, b, 1, -1)$ with

$$a = 2\gamma(\sqrt{2\alpha} - \sqrt{-2\beta} - 1), \qquad b = 2(\sqrt{2\alpha} + \sqrt{-2\beta}).$$

Hence the $x_n(a)$ are solutions of the Painlevé III equation related to this Painlevé V equation with $\delta = 0$. Clarkson [41] in fact used the special function solutions of Painlevé III to give the connection with generalized Charlier polynomials, see also Section 6.2.2.

3.3 Unicity of solutions for d-P_{II}

For the orthogonal polynomials on the unit circle with weight $w(\theta) = e^{t\cos\theta}$ we want a solution of the recurrence relation (3.8), with initial condition $\alpha_{-1} = -1$, for which $-1 < \alpha_n < 1$ for all $n \geq 0$. For the generalized Charlier polynomials with $\beta = 1$ we want a solution of (3.43), with initial value $c_0 = 1$, for which $-1 < c_n < 1$ for all $n \geq 1$, since $a_n^2 = a(1 - c_n^2) > 0$. For both cases we are therefore interested in a solution of

$$x_{n+1} + x_{n-1} = \frac{\alpha n x_n}{1 - x_n^2}, \tag{3.52}$$

with initial condition $x_0 = \pm 1$ and the constraint $-1 < x_n < 1$ for all $n \geq 1$. If we use the recurrence relation (3.52) to compute the sequence by

$$x_{n+1} = \frac{\alpha n x_n}{1 - x_n^2} - x_{n-1},$$

starting from x_0 and $x_1 = I_1(2/\alpha)/I_0(2/\alpha)$ (with $\alpha = 0.2$), then after a few steps we get results outside $[-1, 1]$ (see Figure 3.1).

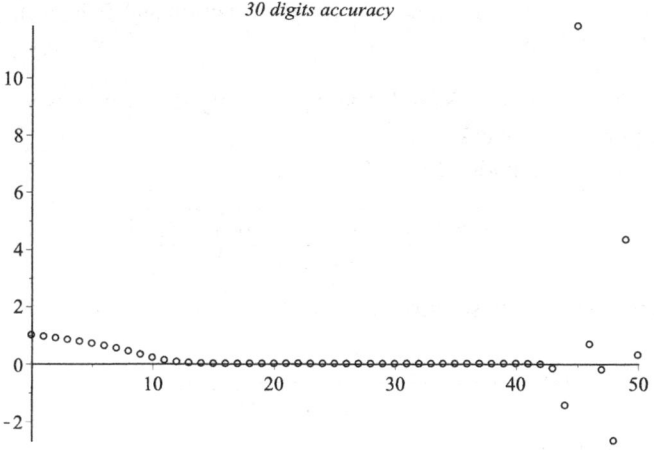

Figure 3.1 Computing x_n from d-P$_{II}$ (with $\alpha = 0.2$) using 30 significant digits

Again the butterfly effect is present: a small change in the initial value x_1, which we cannot compute exactly since it is a ratio of transcendental functions, leads to catastrophic results further on. The reason is that this nonlinear equation has a unique solution for which $x_0 = 1$ and $-1 < x_n < 1$, and this is the solution for which $x_1 = I_1(2/\alpha)/I_0(2/\alpha)$. A small deviation from this initial x_1 hence will give a solution for which $|x_n| > 1$ for some n. Note that $x_n = \pm 1$ for some n will give a singularity for x_{n+1}, which is also not allowed for reflection coefficients or recurrence coefficients of orthogonal polynomials.

Theorem 3.10 *Suppose that $\alpha > 0$. Then there is a unique solution of (3.52) which satisfies $x_0 = 1$ and $-1 < x_n < 1$ for $n \geq 1$. This solution corresponds to the initial condition $x_1 = I_1(2/\alpha)/I_0(2/\alpha)$ and is positive for every $n \geq 0$.*

Similar results hold for modifications of the conditions

Corollary 3.11 *Suppose that $\alpha > 0$. Then there is a unique solution of* (3.52) *which satisfies $x_0 = -1$ and $-1 < x_n < 1$ for $n \geq 1$. This solution corresponds to the initial condition $x_1 = -I_1(2/\alpha)/I_0(2/\alpha)$ and is negative for every $n \geq 0$.*

Proof   The sequence $y_n = -x_n$ satisfies the conditions of Theorem 3.10   □

Corollary 3.12 *Suppose that $\alpha < 0$. Then there is a unique solution of* (3.52) *which satisfies $x_0 = 1$ and $-1 < x_n < 1$ for $n \geq 1$. This solution corresponds to the initial condition $x_1 = I_1(2/\alpha)/I_0(2/\alpha)$.*

Proof   The sequence $y_n = (-1)^n x_n$ satisfies (3.52) but with $-\alpha$ instead of α, hence y_n satisfies the conditions of Theorem 3.10. Observe that $x_1 = -y_1 = -I_1(-2/\alpha)/I_0(-2/\alpha)$, and since I_1 is an odd function and I_0 is even, this gives $x_1 = I_1(2/\alpha)/I_0(2/\alpha)$.   □

Proof of Theorem 3.10   We will give a proof only for the case $\alpha > 1$. The case $\alpha \leq 1$ requires more work.

If we solve the equation (3.52) for x_n, then

$$x_n = \frac{-\alpha n \pm \sqrt{(\alpha n)^2 + 4(x_{n+1} + x_{n-1})^2}}{2(x_{n+1} + x_{n-1})}.$$

Our interest will be the solution with the $+$ sign. If we introduce the function

$$f(t) = \frac{-1 + \sqrt{1 + t^2}}{t} = \frac{t}{1 + \sqrt{1 + t^2}}, \qquad t \in \mathbb{R}, \tag{3.53}$$

then we would have

$$x_n = f\left(\frac{2(x_{n+1} + x_{n-1})}{\alpha n}\right), \qquad n \geq 1. \tag{3.54}$$

Observe that $-1 < f(t) < 1$ for $t \in \mathbb{R}$ and that f is an increasing function with $\lim_{t \to \pm\infty} f(t) = \pm 1$. Furthermore $f(t) > 0$ for $t > 0$.

Let $S = [-1, 1]^{\mathbb{N}_0}$ be the collection of infinite real sequences $x = (x_n)_{n \geq 0}$ with $x_0 = 1$ and $-1 \leq x_n \leq 1$ for $n \geq 1$. Clearly $S \subset \ell_\infty(\mathbb{N})$ with the norm

$$\|x\| = \sup_{n \geq 0} |x_n|.$$

In fact, S is a subset of the unit ball in $\ell_\infty(\mathbb{N})$. We define the mapping $F : S \to S$ by

$$(Fx)_n = \begin{cases} 1, & \text{if } n = 0, \\ f\left(\frac{2(x_{n+1} + x_{n-1})}{\alpha n}\right), & \text{if } n \geq 1, \end{cases} \tag{3.55}$$

where f is given by (3.53). Clearly $-1 < f(t) < 1$ for all $t \in \mathbb{R}$ implies that $Fx \in S$ whenever $x \in S$. Furthermore, for $x, y \in S$

$$(Fx)_n - (Fy)_n = \begin{cases} 0, & \text{if } n = 0, \\ f\left(\frac{2(x_{n+1}+x_{n-1})}{\alpha n}\right) - f\left(\frac{2(y_{n+1}+y_{n-1})}{\alpha n}\right), & \text{if } n \geq 1. \end{cases}$$

If we use the mean value theorem, then

$$f\left(\frac{2(x_{n+1} + x_{n-1})}{\alpha n}\right) - f\left(\frac{2(y_{n+1} + y_{n-1})}{\alpha n}\right)$$
$$= \frac{2f'(\xi_n)}{\alpha n} \left((x_{n+1} - y_{n+1}) + (x_{n-1} - y_{n-1})\right),$$

for some ξ_n between $2(x_{n+1} + x_{n-1})/\alpha n$ and $2(y_{n+1} + y_{n-1})/\alpha n$. Observe that

$$f'(t) = \frac{1}{\sqrt{1 + t^2}(1 + \sqrt{1 + t^2})}$$

so that f' is decreasing on $[0, \infty)$ with $f'(0) = 1/2$ and $\lim_{t \to \infty} f'(t) = 0$, and f' is a symmetric function. This implies that $f'(\xi_n) \leq 1/2$, so that for $n \geq 2$

$$|(Fx)_n - (Fy)_n| \leq \frac{1}{\alpha n} \left(|x_{n+1} - y_{n+1}| + |x_{n-1} - y_{n-1}|\right) \leq \frac{2}{\alpha n} \|x - y\|.$$

Since $x_0 = y_0 = 1$, we have for $n = 1$

$$|(Fx)_1 - (Fy)_1| \leq \frac{1}{\alpha} |x_2 - y_2| \leq \frac{1}{\alpha} \|x - y\|,$$

so that

$$\|Fx - Fy\| \leq \frac{1}{\alpha} \|x - y\|.$$

Hence, if $\alpha > 1$ then F is a contraction on S. The fixed point theorem then implies that there is a unique $x \in S$ for which $Fx = x$, and this x satisfies $x_0 = 1$ and (3.54), so that it is a solution of the discrete Painlevé equation (3.52).

We know that the solution of (3.52) with $x_0 = 1$ and $x_1 = I_1(2/\alpha)/I_0(2/\alpha)$ is such that $a_n^2 = a(1 - x_n^2)$ are recurrence coefficients of generalized Charlier polynomials with $a = 1/\sqrt{\alpha}$ and $\beta = 1$, hence this solution satisfies $-1 < x_n < 1$ and the fixed point solution hence must be the solution with $x_1 = I_1(2/\alpha)/I_0(2/\alpha)$. Furthermore, if we start with a positive sequence x in S then all iterates $F^k x$ are positive, hence the fixed point will be positive as well so that $0 < x_n < 1$ for $n \geq 1$. $\qquad \square$

4

Ladder operators

4.1 Orthogonal polynomials with exponential weights

A useful technique for finding the recurrence coefficients of orthogonal polynomials is the use of ladder operators, i.e., operators that increase the degree of the orthogonal polynomial by one (raising operator) and decrease the degree by one (lowering operator). These ladder operators have a very natural meaning for Hermite polynomials, which appear in the eigenfunctions of the harmonic oscillator. The Schrödinger equation for the harmonic oscillator is

$$-\frac{d^2y}{dx^2} + x^2y = E_n y,$$

where E_n is the energy, x^2 is the potential for the harmonic oscillator and y is the wave function for energy E_n. The Hermite polynomials H_n, for which

$$\int_{-\infty}^{\infty} H_n(x)H_m(x)e^{-x^2}\,dx = 0, \qquad m \neq 0,$$

satisfy the differential equation

$$H_n'' - 2xH_n' + 2nH_n = 0.$$

Putting $y = e^{-x^2/2}H_n$ gives

$$y' = (-xH_n + H_n')e^{-x^2/2}, \quad y'' = (H_n'' - 2xH_n' + x^2H_n - H_n),$$

so that $y'' - x^2y = -(2n + 1)y$, and hence the eigenvalues (energies) for the harmonic oscillator are $E_n = 2n + 1$ and the corresponding eigenfunction is $e^{-x^2/2}H_n(x)$. The Hermite polynomials have the property

$$H_n'(x) = 2nH_{n-1}(x)$$

50

which maps the eigenfunction for E_n to the eigenfunction for E_{n-1} (annihilation or lowering operator). One also has

$$\left(e^{-x^2} H_n(x)\right)' = -e^{-x^2} H_{n+1}(x),$$

which allows to move from energy E_n to E_{n+1} (creation or raising operator). This simple idea was extended to orthogonal polynomials on the real line with a weight function $w(x) = e^{-V(x)}$, for which

$$\int P_n(x)P_m(x)e^{-V(x)} \, dx = h_n \delta_{n,m}.$$

We will use monic orthogonal polynomials $P_n(x) = x^n + \cdots$, in which case one has $h_n = 1/\gamma_n^2$. The function V is often called the potential, and $V(x) = x^2$ corresponds to the Hermite polynomials (and the harmonic oscillator).

Ladder operators for other orthogonal polynomials were known before (one can even go back to Laguerre), but mostly these were obtained case by case. Chen and Ismail [25] found a general setting for ladder operators which contains all the earlier known cases. Recall that the recurrence relation for the monic orthogonal polynomials is

$$xP_n(x) = P_{n+1}(x) + b_n P_n(x) + a_n^2 P_{n-1}(x). \tag{4.1}$$

Theorem 4.1 (Chen–Ismail) *Suppose V is differentiable and V' is Lipschitz continuous on the real line. If $(P_n)_{n\in\mathbb{N}}$ are the monic orthogonal polynomials for the weight $w(x) = e^{-V(x)}$, then*

$$\left(\frac{d}{dz} + B_n(z)\right) P_n(z) = a_n^2 A_n(z) P_{n-1}(z), \tag{4.2}$$

and

$$\left(\frac{d}{dz} - B_n(z) - V'(z)\right) P_{n-1}(z) = -A_{n-1}(z) P_n(z), \tag{4.3}$$

where the functions A_n and B_n are given by

$$A_n(z) = \gamma_n^2 \int \frac{V'(z) - V'(x)}{z - x} P_n^2(x) w(x) \, dx, \tag{4.4}$$

$$B_n(z) = \gamma_{n-1}^2 \int \frac{V'(z) - V'(x)}{z - x} P_{n-1}(x) P_n(x) w(x) \, dx. \tag{4.5}$$

We will prove this result in Section 4.3 using the Riemann–Hilbert problem for orthogonal polynomials. Observe that when V is a polynomial of degree m, then $[V'(z) - V'(x)]/(z - x)$ is a polynomial in z of degree $m - 2$, so that A_n and B_n are polynomials of degree at most $m - 2$ in that case. If V is a rational function, then A_n and B_n are also rational functions but with poles of one order higher.

An important aspect of these ladder relations is that they give two compatibility relations, which often allow to find recurrence relations for the recurrence coefficients of the orthogonal polynomials. The idea is that the ladder relations should be compatible with each other and with the three term recurrence relation for the orthogonal polynomials.

Theorem 4.2 (Chen–Ismail)   *The following two compatibility relations hold between the functions $(A_n, B_n)_n$ and the recurrence coefficients $(a_n^2, b_n)_n$:*

$$B_{n+1}(z) + B_n(z) = (z - b_n)A_n(z) - V'(z), \qquad (4.6)$$

$$a_{n+1}^2 A_{n+1}(z) - a_n^2 A_{n-1}(z) = 1 + (z - b_n)[B_{n+1}(z) - B_n(z)]. \qquad (4.7)$$

Proof   From (4.2) we find

$$P_n'(z) = a_n^2 A_n(z)P_{n-1}(z) - B_n(z)P_n(z),$$

and from (4.3) (with $n \to n + 1$) we find

$$P_n'(z) = [B_{n+1}(z)P_n(z) + V'(z)]P_n(z) - A_n(z)P_{n+1}(z).$$

Eliminating P_n' then gives

$$a_n^2 A_n(z)P_{n-1}(z) = [B_{n+1}(z) + B_n(z) + V'(z)]P_n(z) - A_n(z)P_{n+1}(z).$$

Now use the recurrence relation (4.1) to find

$$(z - b_n)A_n(z)P_n(z) = [B_{n+1}(z) + B_n(z) + V'(z)]P_n(z).$$

If we cancel the common factor $P_n(z)$, which we can do whenever z is not a zero of P_n, then (4.6) follows.

For (4.7) we start from the recurrence relation (4.1), which after taking derivatives gives

$$P_{n+1}'(z) = (z - b_n)P_n'(z) + P_n(z) - a_n^2 P_{n-1}'(z).$$

Rewrite $P_{n+1}'(z)$ and $P_n'(z)$ using (4.2) and $P_{n-1}'(z)$ using (4.3), then one finds

$$a_{n+1}^2 A_{n+1}(z)P_n(z) - B_{n+1}(z)P_{n+1}(z) = (z - b_n)[a_n^2 A_n(z)P_{n-1}(z) - B_n(z)P_n(z)]$$
$$+ P_n(z) + a_n^2[A_{n-1}(z)P_n(z) - B_n(z)P_{n-1}(z) - V'(z)P_{n-1}(z)].$$

Replace $P_{n+1}(z)$ using the three term recurrence relation and collect terms to find

$$\left(a_{n+1}^2 A_{n+1}(z) - a_n^2 A_n(z) - (z - b_n)[B_{n+1}(z) - B_n(z)] - 1\right)P_n(z)$$
$$= a_n^2\left(-B_{n+1}(z) + (z - b_n)A_n(z) - B_n(z) - V'(z)\right)P_{n-1}(z).$$

The right-hand side vanishes because of (4.6), and canceling $P_n(z)$ on the left-hand side then gives the required (4.7).   □

If we multiply (4.7) by $A_n(z)$, and then replace $(z - b_n)A_n(z)$ on the left-hand side by (4.6), then

$$a_{n+1}^2 A_n(z)A_{n+1}(z) - a_n^2 A_{n-1}(z)A_n(z)$$
$$= A_n(z) + [B_{n+1}^2(z) - B_n^2(z)] + V'(z)[B_{n+1}(z) - B_n(z)].$$

This contains differences on both sides of the equation, hence summation and using $a_0^2 = 0$ and $B_0(z) = 0$ gives

$$a_n^2 A_n(z)A_{n-1}(z) = \sum_{j=0}^{n-1} A_j(z) + B_n^2(z) + V'(z)B_n(z). \tag{4.8}$$

4.2 Riemann–Hilbert problem for orthogonal polynomials

The Riemann–Hilbert problem for orthogonal polynomials was formulated by Fokas, Its and Kitaev in 1992 [70]. It describes a boundary value problem that characterizes orthogonal polynomials. Let w be a weight function on the real line which is Hölder continuous. The Riemann–Hilbert problem (RHP) is to find a 2×2 matrix valued function $Y : \mathbb{C} \to \mathbb{C}^{2 \times 2}$ for which the following properties hold:

1. Y is analytic on $\mathbb{C} \setminus \mathbb{R}$;
2. Y has boundary values $Y_\pm(x) = \lim_{\epsilon \to 0+} Y(x \pm i\epsilon)$ that satisfy the *jump condition*

$$Y_+(x) = Y_-(x) \begin{pmatrix} 1 & w(x) \\ 0 & 1 \end{pmatrix}, \qquad x \in \mathbb{R}; \tag{4.9}$$

3. Y has the *asymptotic behavior*

$$Y(z) = \left(\mathbb{I} + O(1/z)\right) \begin{pmatrix} z^n & 0 \\ 0 & z^{-n} \end{pmatrix}, \qquad z \to \infty. \tag{4.10}$$

The (unique) solution to this problem is

$$Y(z) = \begin{pmatrix} P_n(z) & \dfrac{1}{2\pi i} \displaystyle\int_{-\infty}^{\infty} \dfrac{P_n(x)}{x - z} w(x)\,dx \\ -2\pi i \gamma_{n-1}^2 P_{n-1}(z) & -\gamma_{n-1}^2 \displaystyle\int_{-\infty}^{\infty} \dfrac{P_{n-1}(x)}{x - z} w(x)\,dx \end{pmatrix}, \tag{4.11}$$

where $P_n(z) = z^n + \cdots$ is the monic orthogonal polynomial of degree n for the weight function w and

$$\frac{1}{\gamma_{n-1}^2} = \int_{-\infty}^{\infty} P_{n-1}^2(x)w(x)\,dx.$$

Exercise 9: Show that (4.11) is indeed a solution of this Riemann–Hilbert problem. It is sufficient to do this for the first row, since the proof for the second row is very similar. You need to use the *Sokhotsky–Plemelj* formula, which says that the boundary value problem for a function $f : \mathbb{C} \to \mathbb{C}$, which is analytic on $\mathbb{C} \setminus R$ with boundary values $f_+(x) - f_-(x) = w(x)$ for $x \in \mathbb{R}$ and asymptotic behavior $f(z) = O(1/z)$ as $z \to \infty$, is given by the Cauchy transform

$$f(z) = \frac{1}{2\pi i} \int_{-\infty}^{\infty} \frac{w(x)}{x - z} \, dx.$$

To show that this is the unique solution, one considers the scalar function $\det Y$, for which the following hold:

1. $\det Y$ is analytic on $\mathbb{C} \setminus \mathbb{R}$.
2. $\det Y_+(x) = \det Y_-(x)$ for $x \in \mathbb{R}$. This means that $\det Y$ does not have a jump across the real line, and hence $\det Y$ is continuous on $\mathbb{C}$ and hence, by Morera's theorem, $\det Y$ is an entire function.
3. $\det Y(z) = 1 + O(1/z)$ as $z \to \infty$. Hence $\det Y$ is a bounded function and hence, by Liouville's theorem, $\det Y$ is a constant. Taking $z \to \infty$ shows that the constant is 1.

Hence we conclude that $\det Y(z) = 1$, which in view of (4.11) gives the relation

$$P_n(z) \int_{-\infty}^{\infty} \frac{P_{n-1}(x)}{z - x} w(x) \, dx - P_{n-1}(z) \int_{-\infty}^{\infty} \frac{P_n(x)}{z - x} w(x) \, dx = \frac{1}{\gamma_{n-1}^2}. \tag{4.12}$$

This result also follows from the Christoffel–Darboux formula for the orthonormal polynomials $p_n(x) = P_n(x)/\gamma_n$

$$\sum_{k=0}^{n-1} p_k(z) p_k(y) = a_n \frac{p_n(z) p_{n-1}(y) - p_{n-1}(z) p_n(y)}{z - y}$$

by integrating it with respect to $w(y) \, dy$ and using $a_n = \gamma_{n-1}/\gamma_n$.

A consequence of $\det Y = 1$ is that Y^{-1} exists and it is given by

$$Y^{-1} = \begin{pmatrix} -\gamma_{n-1}^2 \int_{-\infty}^{\infty} \frac{P_{n-1}(x)}{x - z} w(x) \, dx & -\frac{1}{2\pi i} \int_{-\infty}^{\infty} \frac{P_n(x)}{x - z} w(x) \, dx \\ 2\pi i \gamma_{n-1}^2 P_{n-1}(z) & P_n(z) \end{pmatrix}. \tag{4.13}$$

Suppose now that $\hat{Y}$ is another solution of the RHP, then one considers the matrix $M(z) = \hat{Y} Y^{-1}$. It satisfies the following RHP:

1. M is analytic in $\mathbb{C} \setminus \mathbb{R}$.
2. $M_+(x) = M_-(x)$, for $x \in \mathbb{R}$.
3. $M(z) = \mathbb{I} + O(1/z)$ as $z \to \infty$.

But then M does not have a jump across the real line, and hence M is an entire matrix function. Liouville's theorem then implies that the entries of M are constant functions, and taking $z \to \infty$ gives $M(z) = \mathbb{I}$, the identity matrix. This means that $\hat{Y} = Y$, and hence the solution of the RHP is unique.

4.3 Proof of the ladder operators

We can now prove Theorem 4.1 using the Riemann–Hilbert problem. Let us consider the matrix function $R = Y'Y^{-1}$, so that $Y' = RY$. The entry $R_{1,1}$ is given by

$$R_{1,1} = (Y')_{1,1}(Y^{-1})_{1,1} + (Y')_{1,2}(Y^{-1})_{2,1},$$

and if we use (4.11) and (4.13), then this gives

$$R_{1,1} = \gamma_{n-1}^2 \left[P_n'(z) \int_{-\infty}^{\infty} \frac{P_{n-1}(x)}{x-z} w(x)\, dx + P_{n-1}(z) \left(\int_{-\infty}^{\infty} \frac{P_n(x)}{x-z} w(x)\, dx \right)' \right].$$

From now on we will use the notation

$$C(f) = \int_{-\infty}^{\infty} \frac{f(x)}{x-z}\, dx$$

for the Cauchy transform of a function f that vanishes sufficiently fast at $\pm\infty$. This Cauchy transform has the following useful properties. If

$$\int_{-\infty}^{\infty} x^k f(x)\, dx = 0, \qquad 0 \le k \le n-1,$$

then for every polynomial π_n of degree $\le n$

$$\pi_n(z)C(f) = C(f\pi_n).$$

Furthermore, if f is differentiable and f' vanishes sufficiently fast at $\pm\infty$, then

$$C(f)' = C(f').$$

The orthogonality thus gives

$$P_n'(z)C(P_{n-1}w) = C(P_n'P_{n-1}w)$$

and

$$P_{n-1}C(P_n w) = C(P_{n-1}P_n w).$$

Differentiating the latter gives

$$P'_{n-1}C(P_nw) + P_{n-1}C(P_nw)' = C(P_{n-1}P_nw)' = C((P_{n-1}P_nw)').$$

Hence

$$R_{1,1} = \gamma_{n-1}^2\left(-C(P_{n-1}P'_nw) + C((P_{n-1}P_nw)') - C(P'_{n-1}P_nw)\right).$$

If we use $w' = -V'w$ then this gives

$$R_{1,1} = -\gamma_{n-1}^2 C(P_{n-1}P_nV'w) = -\gamma_{n-1}^2 \int_{-\infty}^{\infty} P_{n-1}(x)P_n(x)\frac{V'(x)}{x-z}w(x)\,dx.$$

In a similar way one finds

$$R_{1,2} = -\frac{1}{2\pi i}\int_{-\infty}^{\infty} P_n^2(x)\frac{V'(x)}{x-z}w(x)\,dx$$

and

$$R_{2,1} = 2\pi i\gamma_{n-1}^4 \int_{-\infty}^{\infty} P_{n-1}^2(x)\frac{V'(x)}{x-z}w(x)\,dx.$$

For $R_{2,2}$ one also needs to use (4.12) and one obtains

$$R_{2,2} = \gamma_{n-1}^2 \int_{-\infty}^{\infty} P_n(x)P_{n-1}(x)\frac{V'(x)}{x-z}w(x)\,dx.$$

With this information, we now find

$$P'_n(z) = (Y')_{1,1} = R_{1,1}Y_{1,1} + R_{1,2}Y_{2,1}$$
$$= \gamma_{n-1}^2\left(-P_n(z)\int_{\infty}^{\infty} P_n(x)P_{n-1}(x)\frac{V'(x)}{x-z}w(x)\,dx\right.$$
$$\left. + P_{n-1}(z)\int_{-\infty}^{\infty} P_n^2(x)\frac{V'(x)}{x-z}w(x)\,dx\right).$$

This can be written as

$$P'_n(z) = \gamma_{n-1}^2\left(P_{n-1}(z)\int_{-\infty}^{\infty} P_n^2(x)\frac{V'(x) - V'(z)}{x-z}w(x)\,dx\right.$$
$$-P_n(z)\int_{-\infty}^{\infty} P_n(x)P_{n-1}(x)\frac{V'(x) - V'(z)}{x-z}w(x)\,dx\right)$$
$$+ \gamma_{n-1}^2 V'(z)\left(P_{n-1}(z)C(P_n^2w) - P_n(z)C(P_nP_{n-1}w)\right).$$

But $P_{n-1}C(P_n^2w) - P_nC(P_nP_{n-1}w) = P_{n-1}P_nC(P_nw) - P_nP_{n-1}C(P_nw) = 0$, so what is left is precisely (4.2) if we take into account the definitions (4.4)–(4.5).

In a similar way one computes

$$P'_{n-1}(z) = \frac{-1}{2\pi i\gamma_{n-1}^2}(Y')_{2,1} = \frac{-1}{2\pi i\gamma_{n-1}^2}(R_{2,1}Y_{1,1} + R_{2,2}Y_{2,1}),$$

to find (4.3). One needs to use (4.12) to get the term involving $V'(z)$.

4.4 A modification of the Laguerre polynomials

Let's apply these ladder operators to the orthogonal polynomials for the weight function

$$w(x) = x^\alpha e^{-x} e^{-s/x}, \qquad x \in [0, \infty), \tag{4.14}$$

with $\alpha > 0$ and $s \geq 0$. This weight function was studied in detail by Chen and Its [27] and here we shall discuss their findings, but refer to their paper for the detailed calculations. This is the Laguerre weight when $s = 0$, but for $s > 0$ it tends to zero very fast as $x \to 0$. It is a semi-classical weight that satisfies the Pearson equation

$$[x^2 w(x)]' = [s + (\alpha + 2)x - x^2]w(x).$$

The potential is $V(x) = -\log w(x) = x + \frac{s}{x} - \alpha \log x$ for $x \in [0, \infty)$, so that

$$\frac{V'(z) - V'(x)}{z - x} = \frac{1}{z}\left(\frac{\alpha}{x} + \frac{s}{x^2}\right) + \frac{s}{z^2 x}.$$

The functions A_n and B_n in (4.4)–(4.5) (but the integrals are now over $[0, \infty)$ instead of over the real line) are easily found to be

$$A_n(z) = \frac{1}{z} + \frac{c_n}{z^2}, \qquad B_n(z) = -\frac{n}{z} + \frac{d_n}{z^2},$$

where c_n and d_n are given by

$$c_n = s\gamma_n^2 \int_0^\infty \frac{P_n^2(x)}{x} w(x)\,dx, \qquad d_n = s\gamma_{n-1}^2 \int_0^\infty \frac{P_n(x)P_{n-1}(x)}{x} w(x)\,dx.$$

Observe that these depend on the variable s. It is not obvious how to evaluate these integrals, so we consider them as unknown sequences and we will try to find recurrence relations and differential equations for them. The compatibility relation (4.6) is

$$-\frac{2n + 1}{z} + \frac{d_n + d_{n+1}}{z^2} = (z - b_n)\left(\frac{1}{z} + \frac{c_n}{z^2}\right) - 1 + \frac{s}{z^2} + \frac{\alpha}{z}.$$

This is a rational expression with terms $1/z$ and $1/z^2$ only. Comparing the terms for $1/z$ gives

$$b_n = 2n + \alpha + 1 + c_n, \tag{4.15}$$

which expresses the recurrence coefficient b_n in terms of c_n. Observe that for $s = 0$ we have $c_n(0) = 0$ and $b_n(0) = 2n + \alpha + 1$, which is indeed the recurrence coefficient for Laguerre polynomials. Comparing the terms of $1/z^2$ gives

$$d_n + d_{n+1} + b_n c_n = s,$$

which together with (4.15) gives

$$d_n + d_{n+1} = s - [2n + \alpha + 1 + c_n]c_n. \tag{4.16}$$

The second compatibility relation (4.7) becomes

$$a_{n+1}^2 \left(\frac{1}{z} + \frac{c_{n+1}}{z^2} \right) - a_n^2 \left(\frac{1}{z} + \frac{c_{n-1}}{z^2} \right) = 1 + (z - b_n) \left(-\frac{1}{z} + \frac{d_{n+1} - d_n}{z^2} \right).$$

Again, this is a rational expression containing $1/z$ and $1/z^2$. Comparing the terms of $1/z$ gives

$$a_{n+1}^2 - a_n^2 = b_n + d_{n+1} - d_n,$$

and the terms of $1/z^2$ give

$$c_{n+1}a_{n+1}^2 - c_{n-1}a_n^2 = b_n(d_n - d_{n+1}).$$

If we compare the terms of $1/z^2$ in (4.8) then we again find

$$a_n^2 = n(n + \alpha) + d_n + \sum_{j=0}^{n-1} c_j, \tag{4.17}$$

which expresses the recurrence coefficient a_n^2 in terms of the $(c_k)_{0 \le k \le n-1}$ and d_n. Observe that for $s = 0$ we find $a_n^2(0) = n(n + \alpha)$, which is indeed the correct recurrence coefficient for Laguerre polynomials. Comparing the terms of $1/z^3$ in (4.8) gives

$$a_n^2(c_n + c_{n-1}) + (2n + \alpha)d_n = ns,$$

and comparing the terms of $1/z^4$ in (4.8) finally gives

$$a_n^2 c_n c_{n-1} = d_n(d_n - s).$$

Eliminating a_n^2 from the last two equations gives

$$(c_n + c_{n-1})d_n(d_n - s) = [ns - (2n + \alpha)d_n]c_n c_{n-1}. \tag{4.18}$$

The two equations (4.16) and (4.18) give a system of nonlinear recurrence relations for the unknown sequences $(c_n, d_n)_{n \in \mathbb{N}}$. They can be identified as alternate discrete Painlevé II (alt d-P_{II}) equations. If we define $x_n = 1/c_n$ and $y_n = d_n$, then the equations become

$$x_n + x_{n-1} = \frac{ns - (2n + \alpha)y_n}{y_n(y_n - s)}, \tag{4.19}$$

$$y_n + y_{n+1} = s - \frac{2n + \alpha + 1}{x_n} - \frac{1}{x_n^2}, \tag{4.20}$$

which in [96, 8.2.19] corresponds to a discrete Painlevé equation with symmetry the affine Weyl group $(2A_1)^{(1)}$ and surface $D_6^{(1)}$. The initial values that we need are $a_0^2 = 0$, which by (4.17) implies $y_0 = d_0 = 0$, and $b_0 = m_1/m_0$, where the moments are

$$m_k = \int_0^\infty x^{\alpha+k} e^{-x} e^{-s/x} \, dx = 2(\sqrt{s})^{\alpha+k+1} K_{\alpha+k+1}(2\sqrt{s}),$$

where K_ν is the modified Bessel function of the second kind of order ν, which has an integral representation of the form

$$K_\nu(z) = \frac{1}{2}\left(\frac{z}{2}\right)^\nu \int_0^\infty \exp\left(-x - \frac{z^2}{4x}\right) \frac{dx}{x^{\nu+1}},$$

(see, e.g., [123, Eq. 10.32.10]) and for which $K_{-\nu}(z) = K_\nu(z)$. If we use (4.15), then this means that

$$c_0 = b_0 - \alpha - 1 = \frac{\sqrt{s}K_{\alpha+2}(2\sqrt{s}) - (\alpha + 1)K_{\alpha+1}(2\sqrt{s})}{K_{\alpha+1}(2\sqrt{s})} = \sqrt{s}\frac{K_\alpha(2\sqrt{s})}{K_{\alpha+1}(2\sqrt{s})},$$

where the last equality follows from a recurrence relation for the modified Bessel functions. This gives the initial value $x_0 = K_{\alpha+1}(2\sqrt{s})/[\sqrt{s}K_\alpha(2\sqrt{s})]$ for (4.19)–(4.20).

Chen and Its also found a differential equation with respect to the variable s. First they showed that the $c_n(s)$ and $d_n(s)$ satisfy the Toda-type flow

$$sc_n' = 2d_n + (2n + \alpha + 1 + c_n)c_n - s, \tag{4.21}$$

$$sd_n' = \frac{2}{c_n}d_n(d_n - s) + (2n + \alpha + 1)d_n - ns. \tag{4.22}$$

Differentiating (4.21) and then eliminating all the d_n using (4.16), (4.18) and (4.21)–(4.22) gives the differential equation

$$c_n'' = \frac{(c_n')^2}{c_n} - \frac{c_n'}{s} + (2n + \alpha + 1)\frac{c_n^2}{s^2} + \frac{c_n^3}{s^2} + \frac{\alpha}{s} - \frac{1}{c_n}. \tag{4.23}$$

This is very nearly the Painlevé III equation given in (1.20), which after a transformation gives the standard form of Painlevé III.

Exercise 10: Show that the transformation $y(t) = c_n(t^2)/t$ gives the Painlevé III equation (1.20) for y. Identify the parameters.

4.5 Ladder operators for orthogonal polynomials on the linear lattice

The idea for constructing ladder operators as in Theorem 4.1 can be extended. Instead of using the differential operator for orthogonal polynomials with a smooth weight on the real line, one can use the forward difference operator Δ for orthogonal polynomials on a linear lattice, i.e., polynomials satisfying

$$\sum_{k\in\mathbb{Z}} p_m(k)p_n(k)w(k) = \delta_{m,n},$$

where $w(k) \geq 0$ are weights on the linear lattice $\mathbb{Z}$. Recall that the forward and the backward difference operators Δ and ∇ are defined by

$$\Delta f(x) = f(x+1) - f(x), \qquad \nabla f(x) = f(x) - f(x-1).$$

These difference operators are acting on the variable x. Ismail, Nikolova and Simeonov worked out this idea in [92]. The result is

Theorem 4.3 (Ismail, Nikolova and Simeonov) *Define for $x \in \mathbb{Z}$ and $w(x) \neq 0$ the discrete potential as*

$$u(x) = \frac{w(x-1) - w(x)}{w(x)} = -\frac{\nabla w(x)}{w(x)},$$

and let

$$A_n(x) = a_n \sum_{k\in\mathbb{Z}} p_n(k)p_n(k-1)\frac{u(x+1) - u(k)}{x+1-k}w(k),$$

$$B_n(x) = a_n \sum_{k\in\mathbb{Z}} p_n(k)p_{n-1}(k-1)\frac{u(x+1) - u(k)}{x+1-k}w(k).$$

Then one has

$$\Delta p_n(x) = A_n(x)p_{n-1}(x) - B_n(x)p_n(x). \tag{4.24}$$

The functions A_n and B_n are in principle only defined on the integers, but when w satisfies a Pearson equation with difference operator ∇

$$\nabla \sigma(x)w(x) = \tau(x)w(x), \qquad x \in \mathbb{Z},$$

with σ and τ polynomials, then $u(x)$ is a rational function and A_n and B_n will also be rational functions so that the relation (4.24), which is for polynomials, holds for every $x \in \mathbb{C}$. The difference relation (4.24) allows to define a lowering and a raising operator. The lowering operator is $\Delta + B_n(x)$, since

$$\left(\Delta + B_n(x)\right)p_n(x) = A_n(x)p_{n-1}(x).$$

The raising operator is $-\Delta - B_n(x) + A_n(x)(x - b_n)/a_n$ since

$$\left(-\Delta - B_n(x) + \frac{x - b_n}{a_n}A_n(x)\right)p_n(x)$$

$$= A_n(x)\left(-p_{n-1}(x) + \frac{x - b_n}{a_n}p_n(x)\right) = \frac{a_{n+1}}{a_n}A_n(x)p_{n+1}(x),$$

where we used the three term recurrence relation (1.2).

The compatibility of the difference relation (4.24) and the three term recurrence relation (1.2) gives

$$B_n(x) + B_{n+1}(x) = \frac{x - b_n}{a_n}A_n(x) - u(x + 1) + \sum_{j=0}^{n}\frac{A_j(x)}{a_j} \tag{4.25}$$

and

$$a_{n+1}A_{n+1}(x) - a_n^2\frac{A_{n-1}(x)}{a_{n-1}} = (x - b_n)B_{n+1}(x) - (x + 1 - b_n)B_n(x) + 1. \tag{4.26}$$

These can be used to find difference equations for the recurrence coefficients of the orthogonal polynomials and (4.24)–(1.2) are a Lax pair of the resulting difference equation. This particular situation arises for Charlier polynomials (see Section 3.2), generalized Charlier polynomials (Section 3.2.1) and generalized Meixner polynomials (Section 5.3).

4.6 Ladder operators for orthogonal polynomials on a q-lattice

Ladder operators can also be constructed for orthogonal polynomials with weights satisfying a q-difference equation. Chen and Ismail [26] considered orthogonal polynomials on $[0, \infty)$ with a weight function w satisfying a q-difference equation. The potential u is now defined by

$$u(x) = -\frac{D_{q^{-1}}w(x)}{w(x)},$$

where the q-difference operator is

$$D_q f(x) = \begin{cases} \dfrac{f(x) - f(qx)}{x(1-q)}, & \text{if } x \neq 0, \\ f'(0), & \text{if } x = 0. \end{cases}$$

If w satisfies a Pearson equation for the q-difference operator

$$D_{q^{-1}} \sigma(x) w(x) = \tau(x) w(x),$$

with σ and τ polynomials, then u is a rational function. One then has

Theorem 4.4 (Chen–Ismail) *Let*

$$A_n(x) = a_n \int_0^\infty \frac{u(qx) - u(y)}{qx - y} p_n(y) p_n(y/q) w(y) \, dy$$

$$B_n(x) = a_n \int_0^\infty \frac{u(qx) - u(y)}{qx - y} p_n(y) p_{n-1}(y/q) w(y) \, dy,$$

then

$$D_q p_n(x) = A_n(x) p_{n-1}(x) - B_n(x) p_n(x). \tag{4.27}$$

If u is a rational function, then A_n and B_n are also rational functions. The compatibility between the q-difference equation (4.27) and the three term recurrence relation (1.2) now is given by

$$B_n(x) + B_{n+1}(x) = \frac{x - b_n}{a_n} A_n(x) - u(qx) + x(q-1) \sum_{j=0}^n \frac{A_j(x)}{a_j} \tag{4.28}$$

and

$$a_{n+1} A_{n+1}(x) - a_n^2 \frac{A_{n-1}(x)}{a_{n-1}} = (x - b_n) B_{n+1}(x) - (qx - b_n) B_n(x) + 1. \tag{4.29}$$

These will again lead to a nonlinear difference equation for the recurrence coefficients of the orthogonal polynomials. This approach can be used to handle Stieltjes–Wigert polynomials and q-Laguerre polynomials (Section 5.4). These polynomials correspond to a non-determinate moment problem for which there are several measures with the same moments. Some of these measures are absolutely continuous on $[0, \infty)$ but there are also discrete measures on a lattice $\{aq^n : n \in \mathbb{N}\}$, with $q > 1$.

One can also define ladder operators for orthogonal polynomials on a q-lattice $\{aq^n, bq^n : n \in \mathbb{N}\}$, where $0 < q < 1$ or $q > 1$:

$$\int_a^b p_n(x) p_m(x) w(x) \, d_q x = \delta_{m,n},$$

where the q-integral is defined by

$$\int_a^b f(x)\,d_q x = b(1-q)\sum_{k=0}^{\infty} q^k f(bq^k) - a(1-q)\sum_{k=0}^{\infty} q^k f(aq^k).$$

Again the potential u is defined by

$$u(x) = -\frac{D_{q^{-1}} w(x)}{w(x)},$$

and (4.27) holds, but the functions A_n and B_n are now defined by

$$A_n(x) = a_n \int_a^b \frac{u(qx) - u(y)}{qx - y} p_n(y) p_n(y/q) w(y)\,d_q y,$$

$$B_n(x) = a_n \int_a^b \frac{u(qx) - u(y)}{qx - y} p_n(y) p_{n-1}(y/q) w(y)\,d_q y.$$

Some q-discrete Painlevé equations were found for such q-discrete orthogonal polynomials in [13], see Section 5.4.

5

Other semi-classical orthogonal polynomials

In this chapter we will give a few more examples of semi-classical orthogonal polynomials for which the recurrence coefficients satisfy discrete and continuous Painlevé equations. There is absolutely no intention to give a complete inventory of such semi-classical orthogonal polynomials but we restrict ourselves to some typical families. In particular we will also give some Toda evolutions of the classical orthogonal polynomials, some q-orthogonal polynomials (with corresponding q-Painlevé equations) and a general setup for semi-classical orthogonal polynomials on the unit circle. We will not work out all the details anymore but refer to the original paper where these semi-classical families are considered.

5.1 Semi-classical extensions of Laguerre polynomials

Together with my PhD student Lies Boelen we considered in [14] orthogonal polynomials for the weight function

$$w(x) = x^\alpha e^{-x^2+tx}, \qquad x > 0, \tag{5.1}$$

on the semi-infinite interval $[0, \infty)$, with $\alpha > -1$. We showed that the recurrence coefficients a_n^2 and b_n in the three term recurrence relation (1.2) or (1.6) satisfy an asymmetric Painlevé IV equation d-P_{IV} in the following way: put $y_n = 2a_n^2 - n - \alpha/2$ and $x_n = \sqrt{2}/(t - 2b_n)$, then

$$\begin{cases} x_n x_{n-1} = \dfrac{y_n + z_n}{y_n^2 - \frac{\alpha^2}{4}}, \\[2mm] y_n + y_{n+1} = \dfrac{1}{x_n}\left(\dfrac{t}{\sqrt{2}} - \dfrac{1}{x_n}\right), \end{cases} \tag{5.2}$$

where $z_n = n + \alpha/2$. This is a limiting case of the asymmetric Painlevé equation

$$u_n u_{n-1} = \frac{a(v_n + z_n - b)}{v_n^3 - \gamma^2},$$

$$v_n + v_{n+1} = \frac{c}{u_n} + \frac{z_{n+1/2} + d}{u_n - 1}$$

and by taking $u_n = x_n/\epsilon$, $v_n = \epsilon y_n$, $z_n = \epsilon \alpha n + \epsilon \beta$, $a = 1/\epsilon$, $c = 1/\epsilon + t/\sqrt{2}$, $d = -1/\epsilon$ and letting $\epsilon \to 0$ one retrieves (5.2). This asymmetric Painlevé equation for u_n, v_n corresponds to the A_3^c surface in Sakai's classification, see [80, p. 297], or a limiting case of d-P$(D_4^{(1)}/D_4^{(1)})$ in the classification of [96, §8.1.16]. For $t = 0$ the recurrence relations simplify to

$$(y_n + y_{n+1})(y_n + y_{n-1}) = \frac{(y_n^2 - \alpha^2/4)^2}{(y_n + z_n)^2},$$

from which one can obtain a_n^2, and the b_n can be obtained from $2b_n^2 = -y_n - y_{n+1}$, where one takes the positive root to find b_n.

The time evolution in the weight (5.1) corresponds to the Toda lattice from Theorem 3.8. If one eliminates the y_n and y_n' using the Toda equations (3.44)–(3.45), then one finds [65, Eq. (17)]

$$x_n'' = \frac{3}{2}\frac{(x_n')^2}{x_n} + \frac{1}{4}\alpha^2 x_n^3 - \frac{x_n}{8}(t^2 - 4 - 8n - 4\alpha) + \frac{t}{\sqrt{2}} - \frac{3}{4x_n},$$

and if we set $x_n(t) = -\sqrt{2}/q_n(z)$, with $t = 2z$, then one finds the fourth Painlevé equation

$$q_n'' = \frac{(q_n')^2}{2q_n} + \frac{3q_n^3}{2} + 4zq_n^2 + 2(z^2 - 2n - 1 - \alpha)q_n - \frac{2\alpha^2}{q_n},$$

see [65, Thm. 1.1].

5.2 Semi-classical extensions of Jacobi polynomials

The Toda type evolution of Jacobi polynomials was investigated by Basor, Chen and Ehrhardt in [6]. They considered orthogonal polynomials on $[-1, 1]$ with weight function

$$w(x; t) = (1 - x)^\alpha (1 + x)^\beta e^{-tx}, \qquad -1 < x < 1$$

with t a real parameter. For the ladder operators (see Theorem 4.1) one finds as potential

$$V(x) = tx - \alpha \log(1 - x) - \beta \log(1 + x),$$

so that V' is a rational function, and the functions A_n and B_n needed for the ladder operators are rational functions with poles at 1 and -1:

$$A_n(z) = \frac{R_n}{1-z} + \frac{t+R_n}{1+z}, \quad B_n(z) = \frac{r_n}{1-z} + \frac{r_n - n}{1+z},$$

where $R_n = R_n(t)$ and $r_n = r_n(t)$ are functions of t. The compatibility between the three term recurrence relation (1.2) and the lowering relation (4.2) gives (after some straightforward but lengthy calculations) [6, Thm. 4]

$$\begin{cases} 2t(r_{n+1} + r_n) = 4R_n^2 - 2R_n(2n + 1 + \alpha + \beta - t) - 2\alpha t, \\ n(n + \beta) - (2n + \alpha + \beta)r_n = r_n(r_n + \alpha)\left(\frac{t^2}{R_n R_{n-1}} + \frac{t}{R_n} + \frac{t}{R_{n-1}}\right), \end{cases} \quad (5.3)$$

which are the discrete Painlevé equations for this semi-classical weight. The recurrence coefficients a_n, b_n from (1.2) are given in terms of r_n, R_n by [6, Thm. 2]

$$tb_n = 2n + 1 + \alpha + \beta - t - 2R_n,$$

$$t(t + R_n)a_n^2 = n(n + \beta) - (2n + \alpha + \beta)r_n - \frac{tr_n(r_n + \alpha)}{R_n}.$$

The Toda equations (3.44)–(3.45) can be used to find a system of two coupled Riccati equations for r_n and R_n [6, Thm. 3]

$$r\frac{dR_n}{dt} = \alpha t + (2n + \alpha + \beta + 1 - 2t)R_n - 2R_n^2 + 2r_n t,$$

$$-\frac{dr_n}{dt} = \frac{R_n}{t(t + R_n)}\left(n(n + \beta) - (2n + \alpha + \beta)r_n - \frac{t}{R_n}r_n(r_n + \alpha)\right) - \frac{r_n(r_n + \alpha)}{R_n}.$$

Then one can eliminate r_n from the discrete Painlevé equations and the Riccati equations, to find a second order differential equation for R_n. Setting

$$y(t) = 1 + \frac{t}{R_n},$$

one finds the differential equation

$$y'' = \frac{3y - 1}{2y(y - 1)}(y')^2 - \frac{y'}{t} + 2(2n + 1 + \alpha + \beta)\frac{y}{t} - \frac{2y(y + 1)}{y - 1} + \frac{(y - 1)^2}{t^2}\left(\frac{\alpha^2 y}{2} - \frac{\beta^2}{2y}\right).$$

If one puts $Y(t) = y(t/2)$, the differential equation for Y will be the fifth Painlevé equation P_V (see [6, §4]).

5.3 Semi-classical extensions of Meixner polynomials

In Section 3.2.1 we investigated an extension of the Charlier polynomials. Another well-known family of discrete orthogonal polynomials on the integers are the Meixner polynomials, for which the orthogonality measure is the negative binomial or Pascal distribution, with weights

$$w_k = \frac{(\beta)_k c^k}{k!}, \qquad k \in \mathbb{N},$$

where $\beta > 0$ and $0 < c < 1$. The recurrence coefficients in the three term recurrence relation (1.2) for the orthonormal Meixner polynomials are

$$a_n^2 = \frac{n(n+\beta-1)c}{(1-c)^2}, \qquad b_n = \frac{n+(n+\beta)c}{1-c}, \qquad n \in \mathbb{N}.$$

Observe that $a_0^2 = 0$. Together with my student Christophe Smet we investigated in [144] a semi-classical extension of the Meixner polynomials by taking weights

$$w_k = \frac{(\gamma)_k a^k}{(\beta)_k k!}, \qquad \in \mathbb{N}, \tag{5.4}$$

with $\beta, \gamma > 0$. We used the ladder operators on the linear lattice from Theorem 4.3 to find that the recurrence coefficients of these generalized Meixner polynomials are given by

$$a_n^2 = na - (\gamma-1)u_n, \qquad b_n = n + \gamma - \beta + a - \frac{\gamma-1}{a}v_n, \tag{5.5}$$

where (u_n, v_n) satisfy the system on nonlinear equations

$$\begin{cases} (u_n + v_n)(u_{n+1} + v_n) = \dfrac{\gamma-1}{a^2} v_n(v_n - a)\left(v_n - a\dfrac{\gamma-\beta}{\gamma-1}\right), \\[3mm] (u_n + v_n)(u_n + v_{n-1}) = \dfrac{u_n}{u_n - \frac{an}{\gamma-1}}(u_n + a)\left(u_n + a\dfrac{\gamma-\beta}{\gamma-1}\right). \end{cases} \tag{5.6}$$

The initial values for the recurrence coefficients are

$$a_0 = 0, \qquad b_0 = \frac{\gamma a}{\beta}\frac{M(\gamma+1,\beta+1,a)}{M(\gamma,\beta,a)},$$

where $M(a, b, z)$ is the confluent hypergeometric function

$$M(a, b, z) = \sum_{k=0}^{\infty} \frac{(a)_k}{(b)_k k!} z^k = {}_1F_1(a; b; z).$$

The system of recurrence relations (5.6) is a discrete Painlevé system of equations which can be seen as a limiting case of the asymmetric discrete Painlevé IV equation α-d-P$_{\mathrm{IV}}$ given in (1.31), which is d-P($E_6^{(1)}/A_2^{(1)}$) in [96, 8.1.15].

The recurrence coefficients are functions of the parameter a and if we put $a = e^t$, then we have a Toda evolution for the recurrence coefficients as a function of t. The Toda equations (3.44)–(3.45) can then be used to find a second order differential equation for v_n as a function of a. If we put

$$v_n(a) = \frac{a(ay' - (1 + \beta - 2\gamma)y^2 + (1 + n - a + \beta - 2\gamma)y - n)}{2(\gamma - 1)(y - 1)y},$$

then y satisfies the Painlevé V equation

$$y'' = \left(\frac{1}{2y} + \frac{1}{y-1}\right)(y')^2 - \frac{y'}{a} + \frac{(y-1)^2}{a^2}\left(Ay + \frac{B}{y}\right) + \frac{Cy}{a} + \frac{Dy(y+1)}{y-1},$$

where the derivative is with respect to a and the parameters are given by

$$A = \frac{(\beta - 1)^2}{2}, \quad B = -\frac{n^2}{2}, \quad C = n - \beta + 2\gamma, \quad D = -\frac{1}{2}.$$

See [64, Thm. 3] or [12, Thm. 1.2] for the case $\beta = 1$ and Clarkson [41], who used a more direct approach starting from the special function solutions of Painlevé V, see also Section 6.2.4.

In [144] we made the observation that the weights can be written as $w_k = w(k)$, where $w(x)$ is a function of a continuous variable

$$w(x) = \frac{\Gamma(\beta)}{\Gamma(\gamma)} \frac{\Gamma(\gamma + x)a^x}{\Gamma(\beta + x)\Gamma(x + 1)},$$

and that $w(x) = 0$ for $x = -1, -2, -3, \ldots$, giving weights which are non-zero on the integers $\mathbb{N} = \{0, 1, 2 \ldots\}$, but that one also has $w(x) = 0$ whenever $\Gamma(\beta + x)$ has a pole, hence for $x = -\beta, -\beta - 1, -\beta - 2, \ldots$, so that one also has positive weights $w(k + \beta - 1)$ for $k \in \mathbb{N}$, when $\beta < 2$. So one can also investigate orthogonal polynomials $(q_n)_n$ on the shifted lattice $\mathbb{N} + 1 - \beta$:

$$\sum_{k=0}^{\infty} q_n(k + 1 - \beta)q_m(k + 1 - \beta)w(k + 1 - \beta) = \delta_{n,m}.$$

The proof using the compatibility of the three term recurrence relation (1.2) with the lowering relation (4.24) gives that the recurrence coefficients are again given by (5.5), where (u_n, v_n) satisfy the nonlinear equations (5.6), but with a different initial value for b_0:

$$a_0^2 = 0, \quad b_0 = (1 - \beta)\frac{M(\gamma - \beta + 1, 1 - \beta, a)}{M(\gamma - \beta + 1, 2 - \beta, a)}.$$

One can also take a linear combination of the measure μ_1 supported on $\mathbb{N}$ and the measure μ_2 supported on $\mathbb{N} + 1 - \beta$. If $0 < \beta < 2$ and if we take $\mu = c_1\mu_1 + c_2\mu_2$, with $c_1, c_2 > 0$, then one deals with a positive measure on the union of two lattices. The corresponding orthogonal polynomials again satisfy a three term recurrence relation as in (1.2) and the recurrence coefficients are again given (5.5), with (u_n, v_n) satisfying the discrete Painlevé equations (5.6). The only difference is in the initial conditions, which are now

$$a_0^2 = 0, \quad b_0 = \frac{c_1 m_1 + c_2 \hat{m}_1}{c_1 m_0 + c_2 \hat{m}_0},$$

where m_0 and m_1 are the first two moments of μ_1 and $\hat{m}_0$ and $\hat{m}_1$ the first two moments of μ_2, which can be expressed in terms of the confluent hypergeometric function. Observe that b_0 depends on one parameter $\lambda = c_1/c_2 > 0$ and that for $\lambda \to 0$ we get the orthogonal polynomials on the shifted lattice $\mathbb{N} + 1 - \beta$ and for $\lambda \to \infty$ we get the orthogonal polynomials on $\mathbb{N}$. The behavior of the recurrence coefficients $(a_n^2, b_n)_n$ is quite different for the extreme cases $\lambda = 0$ and $\lambda = \infty$, in which case the recurrence coefficients have a nice monotonic behavior. For $0 < \lambda < \infty$ the behavior is oscillatory, reflecting the fact that the support of the measure consists of two lattices. If we take $a = 3, \beta = 2/3$, $\gamma = 9/10$ and $\lambda = 2$ then the recurrence coefficients $(a_n)_n$ and $(b_n)_n$ are plotted in Figure 5.1. Note that the black curve (for the lattice $\mathbb{N}$) and the gray curve (for the shifted lattice $\mathbb{N}+1-\beta$) are nearly the same, but the wavy curve (for the bi-lattice $\mathbb{N} \cup (\mathbb{N} + 1 - \beta)$) is much different. Using results for discrete orthogonal polynomials (e.g., [105]) combined with the discrete Painlevé equations (5.6) one can show that for the lattice $\mathbb{N}$ one has the asymptotic behavior

$$\lim_{n\to\infty}(a_n^2 - an) = (\gamma - \beta)a, \quad \lim_{n\to\infty}(b_n - n) = a.$$

For the shifted lattice $\mathbb{N} + 1 - \beta$ the asymptotic behavior of the recurrence coefficients is

$$\lim_{n\to\infty}(a_n^2 - an) = (\gamma - 1)a, \quad \lim_{n\to\infty}(b_n - n) = a + 1 - \beta.$$

Observe that for $\beta = \gamma$ the weights on $\mathbb{N}$ are the weights for Charlier polynomials, for which $a_n^2 = an$ and $b_n = n + a$, whereas for $\gamma = 1$ the weights on $\mathbb{N} + 1 - \beta$ are the weights for Charlier polynomials shifted by $1 - \beta$, hence $a_n^2 = an$ and $b_n = n + a + 1 - \beta$, which is consistent with the asymptotic behavior given above. If we combine both lattices, then the asymptotic behavior is different and we conjecture that

$$a_n^2/n^{3/2} = O(1), \quad b_n - \frac{n}{2} = O(1).$$

This idea of using a shifted lattice and a bi-lattice can also be used for the generalized Charlier polynomials in Section 3.2.1.

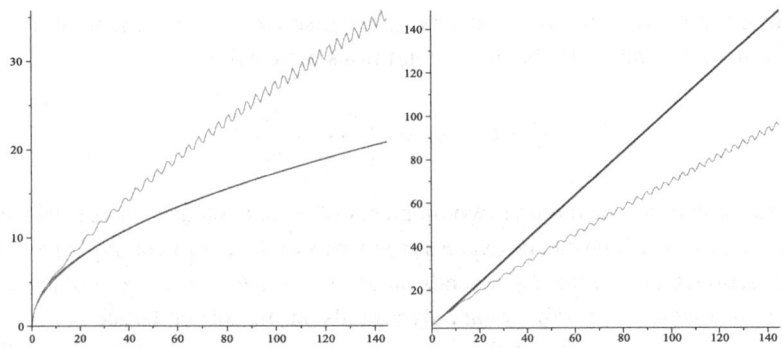

Figure 5.1 The recurrence coefficients a_n (left) and b_n (right) for generalized Meixner polynomials for the lattice $\mathbb{N}$ (black), $\mathbb{N} + 1 - \beta$ (gray) and the bi-lattice (wavy)

5.4 Semi-classical extensions of Stieltjes–Wigert and q-Laguerre polynomials

One can also handle semi-classical weights for the q-difference operator. In [15] we investigated semi-classical extensions of the Stieltjes–Wigert polynomials, the q-Laguerre polynomials and the little q-Laguerre polynomials. The Stieltjes–Wigert weight

$$w_1(x) = \exp(-\log^2 x), \qquad x \in [0, \infty),$$

is the classical example of an indeterminate weight, given by Stieltjes. There exist several measures with the same moments, some of them are absolutely continuous, others are discrete. Wigert generalized this to

$$w_k(x) = \exp(-k^2 \log^2 x), \qquad x \in [0, \infty),$$

for which the moments are

$$m_n = \frac{\sqrt{\pi}}{k} q^{-(n+1)^2/2}, \qquad q = \exp\left(-\frac{1}{2k^2}\right).$$

Askey gave another weight

$$\tilde{w}_\alpha = \frac{x^\alpha}{(-x; q)_\infty (-q/x; q)_\infty},$$

which for $\alpha = 1/2$ has the same moments as w_k. Here we used the q-Pochhammer symbol

$$(a; q)_n = \prod_{k=0}^{n-1} (1 - aq^k), \qquad (a; q)_\infty = \prod_{k=0}^{\infty} (1 - aq^k), \tag{5.7}$$

where q is such that $0 < q < 1$, so that the infinite product converges. The recurrence coefficients for the corresponding Stieltjes–Wigert polynomials are

$$a_n^2 = q^{-4n}(1 - q^n), \quad b_n = q^{-2n-3/2}(1 + q - q^{n+1}),$$

and hence they grow exponentially fast. A semi-classical extension of the Stieltjes–Wigert weight is

$$w(x) = \frac{x^\alpha}{(-x^2; q^2)_\infty(-q^2/x^2; q^2)_\infty}, \quad x \in [0, \infty),$$

where we basically changed x to x^2 and q to q^2. Using the ladder operators for the q-difference operator from Section 4.6, we showed in [15, Thm. 1.1] that the recurrence coefficients are given by

$$a_n^2 = q^{1-n}x_n + q^{-2n-\alpha+1},$$
$$b_n^2 q^{2n+\alpha}x_n = x_{n+1} + q^{2n+2\alpha}x_{n-1}(x_n + q^{-n-\alpha})^2 + 2(x_n + q^{-\alpha}),$$

where $(x_n)_{n\in\mathbb{N}}$ satisfies the recurrence

$$x_{n-1}x_{n+1} = \frac{(x_n + q^{-\alpha})^2}{(q^{n+\alpha}x_n + 1)^2}, \tag{5.8}$$

which is known as a q-discrete Painlevé III equation (q-$\mathrm{P}_{\mathrm{III}}$). The initial values for the required solution are $x_0 = -q^{-\alpha}$ and $x_1 = b_0^2 = (m_1/m_0)^2$, where m_0 and m_1 are the first two moments of w.

A closely related family of orthogonal polynomials with indeterminate weight are the q-Laguerre polynomials, for which the weight function is

$$v_1(x) = (\frac{-p}{\sqrt{q}x}; q)_\infty w_k(x), \quad x \in [0, \infty),$$

with $0 < p < 1$ an extra parameter (for $p = 0$ one gets the Stieltjes–Wigert weight). The recurrence coefficients for the q-Laguerre polynomials are

$$a_n^2 = q^{-4n}(1 - q^n)(1 - pq^{n-1}), \quad b_n = q^{-n-3/2}(-p - q + (1 + q)q^{-n}).$$

If we write $p = q^{\alpha+1}$, then another weight with the same moments is

$$v_2(x) = \frac{x^\alpha}{(-x; q)_\infty}.$$

A semi-classical extension of these weights is

$$w(x) = \frac{x^\alpha(-p/x^2; q^2)_\infty}{(-x^2; q^2)_\infty(-q^2/x^2; q^2)_\infty}, \quad x \in [0, \infty),$$

with $p \in [0, q^{-\alpha})$ and $\alpha \geq 0$. The q-ladder operators allowed us to find [15, Thm. 1.2] that

$$a_n^2 = q^{-n+1} z_n \sqrt{pq^{-2-\alpha}} + q^{-2n-\alpha+1},$$

$$b_n^2 q^{2n+\alpha} z_n^2 = z_n z_{n+1} - 1 + q_{2n+2\alpha}(\sqrt{pq^{-2-\alpha}} z_n + q^{-n-\alpha})^2 (z_n z_{n-1} - 1)$$

$$+ 2(z_n + \sqrt{q^{2-\alpha}/p})(z_n + \sqrt{pq^{\alpha-2}}),$$

where $(z_n)_{n \in \mathbb{N}}$ satisfies the recurrence relation

$$(z_n z_{n-1} - 1)(z_n z_{n+1} - 1) = \frac{(z_n + \sqrt{q^{2-\alpha}/p})^2 (z_n + \sqrt{pq^{\alpha-2}})^2}{(q^{n+\alpha/2-1} \sqrt{p} z_n + 1)^2}. \tag{5.9}$$

This is known as a q-discrete Painlevé V equation (q-P_V). The initial values for the required solution are

$$z_0 = -\sqrt{\frac{q^{2-\alpha}}{p}}, \qquad z_1 = \frac{m_2 m_0 - m_1^2 - m_0^2 q^{-\alpha-1}}{m_0^2 \sqrt{pq^{-\alpha-2}}},$$

where m_0, m_1, m_2 are the first three moments of this weight.

We also investigated a semi-classical extension of the little q-Laguerre polynomials. These little q-Laguerre polynomials $P_n(x|q)$ are orthogonal on the q-lattice

$$\sum_{k=0}^{\infty} P_n(q^k|q) P_m(q^k|q) q^k \frac{q^{k\alpha}}{(q;q)_k} = 0, \qquad m \neq n,$$

and this can also be written in terms of Jackson's q-integral

$$\int_0^1 P_n(x|q) P_m(x|q) x^\alpha (qx;q)_\infty d_q x = 0, \qquad m \neq n,$$

where for $0 < q < 1$ the q-integral is defined as

$$\int_0^1 f(x) d_q x = (1-q) \sum_{k=0}^{\infty} q^k f(q^k).$$

The recurrence coefficients of the little q-Laguerre polynomials are

$$a_n^2 = q^{2n+\alpha-1}(1-q^n)(1-q^{n+\alpha}), \qquad b_n = q^n(1 + q^\alpha - q^{n+\alpha}(1+q)).$$

Observe that these recurrence coefficients are bounded and tend to zero exponentially fast. Changing q to $1/q$ and changing the sign for b_n gives the q-Laguerre polynomials. A semi-classical extension for these little q-Laguerre polynomials uses the weight

$$w(x) = x^\alpha (q^2 x^2; q^2)_\infty,$$

on the q-lattice $\{q^k, k = 0, 1, 2, \ldots\}$, with $\alpha > 0$. The q-ladder operators and the q-lattice then allowed us to find in [15, Thm. 1.3] that the recurrence coefficients are given by

$$a_n^2 = q^{n+\alpha/2-1}(x_n - q^{n+\alpha/2}),$$
$$b_n^2 q^{-2n-\alpha} x_n^2 = 1 - x_n x_{n+1} - q^{-2n}(x_n x_{n-1} - 1)(x_n q^{-\alpha/2} - q^n)^2$$
$$- 2(x_n - q^{\alpha/2})(x_n - q^{-\alpha/2}),$$

and $b_0^2 = 1 - q^{\alpha/2}x_1$, where the $(x_n)_{n \in \mathbb{N}}$ satisfy the recurrence relation

$$(x_n x_{n-1} - 1)(x_n x_{n+1} - 1) = \frac{q^{2n+\alpha}(x_n - q^{\alpha/2})^2 (x_n - q^{-\alpha/2})^2}{(x_n - q^{n+\alpha/2})^2}, \qquad (5.10)$$

and initial values $x_0 = q^{\alpha/2}$ and $x_1 = q^{-\alpha/2}(1 - m_1^2/m_0^2)$, with m_0 and m_1 the first two moments of this semi-classical weight. Again we have the q-discrete Painlevé V equation.

One of the q-extensions of the Hermite polynomials are the q-discrete Hermite polynomials satisfying

$$\int_{-1}^{1} p_n(x) p_m(x) w(x) \, d_q x = \delta_{m,n},$$

with $w(x) = (q^2 x^2; q^2)_\infty$. The recurrence coefficients are

$$a_n^2 = q^{n-1}(1 - q^n), \quad b_n = 0.$$

A number of semi-classical extensions are possible, giving q-extensions for the Freud weights from Section 2.1. The most obvious one is to take

$$w(x) = (q^4 x^4; q^4)_\infty, \qquad x \in [-1, 1]$$

in the q-integral. The recurrence coefficients are then given by

$$a_n^2 = q^{n-1} y_n, \quad b_n = 0,$$

where $(y_n)_{n \in \mathbb{N}}$ satisfies the recurrence relation

$$1 - y_n^2 = q^n(y_{n+1} y_n + 1)(y_{n-1} y_n + 1).$$

A more general weight is

$$w(x) = |x|^\alpha (x^2 q^2; q^2)_\infty (c x^2 q^2; q^2)_\infty, \qquad c \le 1$$

(for $\alpha = 0$ and $c = -1$ one retrieves the earlier weight). It was shown in [13] that the recurrence coefficients are given by $b_n = 0$ and

$$a_{2n}^2 = q^{2n-1+\alpha} u_n, \quad a_{2n+1}^2 = q^{2n} v_n/c,$$

where $(u_n, v_n)_{n\in\mathbb{N}}$ satisfy the system of recurrence relations

$$\begin{cases} q^{2n}(1 - u_n v_n)(1 - u_n v_{n-1}) = (1 - u_n)(1 - c u_n), \\ q^{2n+\alpha+1}(1 - u_n v_n)(1 - u_{n+1} v_n) = (1 - v_n)(1 - v_n/c). \end{cases} \quad (5.11)$$

These are a limiting case of the asymmetric q-discrete Painlevé V equation (α-q-P_V (related to E_6^q in Sakai's classification)

$$\begin{cases} (1 - u_n v_n)(1 - u_n v_{n-1}) = \dfrac{(u_n - 1/p)(u_n - 1/r)(u_n - 1/s)(u_n - 1/t)}{(u_n - b\rho_n)(u_n - \rho_n/b)} \\ (1 - u_n v_n)(1 - u_{n+1} v_n) = \dfrac{(v_n - p)(v_n - r)(v_n - s)(v_n - t)}{(v_n - a\omega_n)(v_n - \omega_n/a)} \end{cases}$$

$$(5.12)$$

with $p = 1, r = c, s = \kappa, t = 1/c\kappa, a = \kappa, b = c\kappa, \rho_n = q^{2n}, \omega_n = q^{2n+\alpha+1}$ and $\kappa \to 0$.

5.5 Semi-classical bi-orthogonal polynomials on the unit circle

Forrester and Witte have worked out the theory for bi-orthogonal polynomials on the unit circle [75]. These bi-orthogonal polynomials $(\phi_n, \hat{\phi}_n)_{n\in\mathbb{N}}$ satisfy the bi-orthogonality relations

$$\frac{1}{2\pi i} \int_{\mathbb{T}} \phi_n(z)\hat{\phi}_m(\bar{z})w(z)\,dz = \delta_{n,m},$$

where $\mathbb{T} = \{z \in \mathbb{C} : |z| = 1\}$ is the unit circle, $z = e^{i\theta}, \bar{z} = e^{-i\theta}$ and w is a semi-classical weight of the form

$$w(z) = \sum_{j=1}^{m} (z - z_j)^{\rho_j}, \qquad \rho_j \in \mathbb{C},$$

with singularities or zeros at m isolated points z_j, which they allow to depend on an extra parameter t. Such weights are also known as generalized Jacobi weights. This weight satisfies the Pearson-type equation

$$w'(z) = \frac{2V(z)}{W(z)}w(z),$$

where V and W are polynomials with $\deg V < m$ and $\deg W = m$. If the weight w is real and positive on the unit circle, then the bi-orthogonal polynomials are orthogonal polynomials on the unit circle together with their complex conjugates, i.e., $\hat{\phi}_n(\bar{z}) = \overline{\phi_n(z)}$. A lot of the theory of orthogonal polynomials on the unit circle also holds for these bi-orthogonal polynomials. In particular there is a recurrence relation like the Szegő recurrence (3.2)

$$\kappa_n \phi_{n+1}(z) = \kappa_{n+1} z \phi_n(z) + \phi_{n+1}(0) \phi_n^*(z), \tag{5.13}$$

$$\kappa_n \phi_{n+1}^*(z) = \kappa_{n+1} \phi_n^*(z) + \hat{\phi}_{n+1}(0) z \phi_n(z), \tag{5.14}$$

where $\kappa_n > 0$ is the leading coefficient of ϕ_n and $\hat{\phi}_n$, and $\phi_n^*(z) = z^n \hat{\phi}_n(1/z)$ is obtained from $\hat{\phi}_n$ by reversing the coefficients. This recurrence relation can be written as

$$\begin{pmatrix} \phi_{n+1}(z) \\ \phi_{n+1}^*(z) \end{pmatrix} = B_n(z) \begin{pmatrix} \phi_n(z) \\ \phi_n^*(z) \end{pmatrix} \tag{5.15}$$

with the transition matrix

$$B_n = \frac{1}{\kappa_n} \begin{pmatrix} \kappa_{n+1} z & \phi_{n+1}(0) \\ \hat{\phi}_{n+1}(0) z & \kappa_{n+1} \end{pmatrix}.$$

Forrester and Witte also worked out the corresponding ladder operators and they obtained [75, Prop. 2.6]

$$W(z) \phi_n'(z) = \Theta_n(z) \phi_{n+1}(z) - (\Omega_n(z) + V(z)) \phi_n(z),$$

$$W(z) \hat{\phi}_n^{*'}(z) = -\Theta_n^*(z) \phi_{n+1}^*(z) + (\Omega_n^*(z) - V(z)) \phi_n^*(z),$$

where Θ_n, Θ_n^* are polynomials of degree $\leq m - 2$ and Ω_n, Ω_n^* are polynomials of degree $\leq m - 1$. These relations and the recurrence relations (5.13)–(5.14) can be written as

$$W(z) \begin{pmatrix} \phi_n(z) \\ \phi_n^*(z) \end{pmatrix}' = A_n(z) \begin{pmatrix} \phi_n(z) \\ \phi_n^*(z) \end{pmatrix} \tag{5.16}$$

with the matrix

$$A_n(z) = \begin{pmatrix} -\left(\Omega_n(z) + V(z) - \dfrac{\kappa_{n+1}}{\kappa_n} z \Theta_n(z) \right) & \dfrac{\phi_{n+1}(0)}{\kappa_n} \Theta_n(z) \\ -\dfrac{\hat{\phi}_{n+1}(0)}{\kappa_n} z \Theta_n^*(z) & \Omega_n^* - V(z) - \dfrac{\kappa_{n+1}}{\kappa_n} \Theta_n^*(z) \end{pmatrix}.$$

The compatibility between (5.15) and (5.16) gives the matrix equation

$$B_n'(z) = A_{n+1}(z) B_n(z) - B_n(z) A_n(z).$$

This compatibility relation gives various linear recurrences for the Θ_n, Θ_n^* and Ω_n, Ω_n^*, and by matching the coefficients of the powers of z for these polynomials, one gets nonlinear recurrences for some of the unknown coefficients of the bi-orthogonal polynomials. In particular, if we set

$$\phi_n(z) = \kappa_n z^n + \ell_n z^{n-1} + \cdots + \phi_n(0),$$

$$\hat{\phi}_n(z) = \kappa_n z^n + \hat{\ell}_n z^{n-1} + \cdots + \hat{\phi}_n(0),$$

then there are nonlinear relations for the reflection coefficients $r_n = \phi_n(0)/\kappa_n$, $\hat{r}_n = \hat{\phi}_n(0)/\kappa_n$ and the subleading coefficients $\beta_n = \ell_n/\kappa_n$ and $\hat{\beta}_n = \hat{\ell}_n/\kappa_n$ of the monic bi-orthogonal polynomials. Comparing the coefficients of z^n in (5.13) gives $\beta_{n+1} - \beta_n = r_{n+1}\hat{r}_n$ so that

$$\beta_n = \sum_{k=1}^n r_k\hat{r}_{k-1}.$$

Forrester and Witte [74] worked this out in more detail for the case $m = 3$ and the weight

$$w(z) = z^{-\mu-\omega}(1+z)^{2\omega_1}(1+tz)^{2\mu}\begin{cases} 1, & \theta \notin (\pi-\phi,\pi), \\ 1-\xi, & \theta \in (\pi-\phi,\pi), \end{cases} \tag{5.17}$$

where μ, ω_1, ω_2 are complex parameters ($\omega = \omega_1 + i\omega_2$), and $t = e^{i\phi}$ and ξ are complex variables. They found the discrete Painlevé equations [74, Prop. 3.4]

$$g_{n+1}g_n = t\frac{(f_n+n)(f_n+n+2\mu)}{f_n(f_n-2\omega_1)}, \tag{5.18}$$

$$f_n + f_{n-1} = 2\omega_1 + \frac{n-1+\mu+\omega}{g_n-1} + \frac{(n+\mu+\bar{\omega})t}{g_n-t}, \tag{5.19}$$

where g_n and f_n are given in terms of the r_n, β_n as

$$g_n = t\frac{(n-1+\mu+\omega)r_{n-1} + (n+\mu+\bar{\omega})r_n}{(n-1+\mu+\omega)r_{n-1} + (n+\mu+\bar{\omega})tr_n},$$

$$f_n = \frac{1}{1-t}\left(t\beta_n - n - (n+1+\mu+\bar{\omega})(1-r_n\hat{r}_n)t\frac{r_{n+1}}{r_n}\right),$$

and a similar system for $\hat{g}_n$ and $\hat{f}_n$ involving $\hat{r}_n$ and $\hat{\beta}_n$. The initial conditions are

$$f_0 = 0, \quad g_1 = t\frac{\mu+\omega+(1+\mu+\bar{\omega})r_1}{\mu+\omega+(1+\mu+\bar{\omega})tr_1},$$

which corresponds to $r_0 = 1$ and $\beta_0 = 0$. The system (5.18)–(5.19) is known as discrete Painlevé V (d-P_V) and corresponds to Sakai's surface $D_4^{(1)}$ and symmetry $D_4^{(1)}$ [96, Eq. 8.1.16]. This discrete Painlevé equation is related to Bäcklund transformations of Painlevé VI.

Exercise 11: In [115] Magnus investigated the weight function $(0 < \alpha < 1)$

$$w(e^{i\theta}) = \begin{cases} A, & \alpha\pi < \theta < 2\pi - \alpha\pi, \\ B, & -\alpha\pi < \theta < \alpha\pi, \end{cases} \tag{5.20}$$

with $A, B \geq 0$ and $A + B > 0$. Show that for this weight one has

$$(n+1)r_{n+1} + (n-1)r_{n-1} = \frac{2r_n}{1-r_n^2}(\beta_n + n\cos\alpha\pi), \tag{5.21}$$

where $\beta_n = \sum_{j=1}^{n} r_j r_{j-1}$, with initial conditions $r_0 = 1, \beta_0 = 0$ and

$$r_1 = \frac{\sin\alpha\pi}{\pi} \frac{A - B}{(1-\alpha)A + \alpha B}.$$

Note that this corresponds to the case $\mu = 0 = \omega$ of (5.17) after a rotation on the unit circle.

A q-version of this analysis was worked out by Biane [8] who considered the real and positive weight

$$w(e^{i\theta}) = \left| \frac{(ae^{i\theta}; q)_\infty}{(be^{i\theta}; q)_\infty} \right|^2, \qquad a, b \in \mathbb{C},$$

on the unit circle, with $0 < q < 1$. Here $(a; q)_\infty$ is the q-Pochhammer symbol given in (5.7). This weight, also known as a q-gamma weight, can also be written as

$$w(z) = \frac{(az; q)_\infty (\bar{a}/z; q)_\infty}{(bz; q)_\infty (\bar{b}/z; q)_\infty},$$

and it satisfies the Pearson-type equation

$$w(qz) = \frac{V(z)}{W(z)} w(z),$$

with polynomials V and W given by

$$V(z) = (qz - \bar{a})(1 - bz), \qquad W(z) = (qz - \bar{b})(1 - az).$$

The case $a = \sqrt{q}, b = 0$ corresponds to the Rogers–Szegő orthogonal polynomials on the unit circle, for which the reflection coefficients are $r_n = q^{n/2}$. The case $a = \sqrt{q}, b = c\sqrt{q}$ corresponds to orthogonal polynomials introduced by Askey [123, Eq. 18.33.15–16], for which the reflection coefficients are

$$r_n = \frac{1-c}{1-cq^n} q^{n/2}.$$

For these two special cases the polynomials V and W have a common factor, and they can be considered as classical weights on the unit circle. The general case with $a, b \in \mathbb{C}$ and $a \neq \sqrt{q}$ and $a \neq b$ gives a semi-classical weight on the unit circle. One can write the Szegő recurrence relations (3.2) for the monic orthogonal polynomials in matrix form as

$$\begin{pmatrix} \Phi_{n+1}(z) \\ \Phi_{n+1}^*(z) \end{pmatrix} = B_n(z) \begin{pmatrix} \Phi_n(z) \\ \Phi_n^*(z) \end{pmatrix},$$

with transition matrix

$$B_n(z) = \begin{pmatrix} z & r_{n+1} \\ z\bar{r}_{n+1} & 1 \end{pmatrix},$$

with $r_n = \Phi_n(0)$. Note that $\alpha_n = -\bar{r}_{n+1}$ are the Verblunsky coefficients as in Section 3.1. The q-ladder operators can be written as

$$V(z) \begin{pmatrix} \Phi_n(qz) \\ \Phi_n^*(qz) \end{pmatrix} = A_n(z) \begin{pmatrix} \Phi_n(z) \\ \Phi_n^*(z) \end{pmatrix},$$

with

$$A_n(z) = \begin{pmatrix} \Omega_n(z) - z\Theta_n(z) & -r_{n+1}\Theta_n(z) \\ -z\bar{r}_{n+1}\Theta_n^*(z) & \Omega_n^*(z) - \Theta_n^*(z) \end{pmatrix},$$

where Θ_n, Θ_n^* and Ω_n, Ω_n^* are polynomials given by

$$\Theta_n(z) = -(a - bq^{n+1})z + (\bar{a} - \bar{b}q^n)\frac{r_n}{r_{n+1}},$$

$$\Theta_n^*(z) = -q(a - bq^n)\frac{\bar{r}_n}{\bar{r}_{n+1}}z + \bar{a} - \bar{b}q^{n+1},$$

$$\Omega_n(z) = z\Theta_n(z) - bq^{n+1}z^2 + t_n z - \bar{b}q^n,$$

$$\Omega_n^*(z) = \Theta_n^*(z) - aqz^2 + \hat{\imath}_n z - \bar{a},$$

with

$$t_n = -r_{n+1}\bar{r}_n(a - bq^{n+1}) - b\beta_n q^n(1 - q) + (\bar{a}b + q)q^n,$$

$$\hat{\imath}_n = r_n\bar{r}_{n+1}(\bar{a} - \bar{b}q^{n+1}) + \bar{a}\bar{\beta}_n(1 - q) + \bar{a}b + q.$$

The compatibility relations are

$$A_{n+1}(z)B_n(z) = B_n(qz)A_n(z),$$

and if we compare the matrix elements and equate the coefficients of equal powers of z, then one finds nonlinear recurrence relations for r_n and β_n:

$$(1 - |r_{n+1}|^2)\big(r_{n+2}(a - bq^{n+2}) + r_n q(\bar{a} - \bar{b}q^n)\big)$$
$$= r_{n+1}(1 - q)[\bar{a}\bar{\beta}_{n+1} + bq^{n+1}\beta_{n+1}] + r_{n+1}(1 - q^{n+1})(\bar{a}b + q),$$

and

$$(1 - |r_{n+1}|^2)\big(\bar{r}_{n+2}(\bar{a} - \bar{b}q^{n+2}) + \bar{r}_n q(a - bq^n)\big)$$
$$= \bar{r}_{n+1}(1 - q)[\bar{a}\bar{\beta}_{n+1} + bq^{n+1}\beta_{n+1}] + \bar{r}_{n+1}(1 - q^{n+1})(\bar{a}b + q).$$

If a and b are real, then $w(e^{-i\theta}) = w(e^{i\theta})$ so that the reflection coefficients r_n and the coefficients β_n are real. The two recurrence relations then reduce to one recurrence relation

$$r_{n+2}(a - bq^{n+2}) + qr_n(a - bq^n) = \frac{2r_{n+1}}{1 - r_{n+1}^2}\big((1-q)(a + bq^{n+1})\beta_{n+1} + (1 - q^{n+1})(ab + q)\big).$$

Notice the similarity with d-P_{II} from (1.25), except for the extra β_{n+1} in the right-hand side and the q-nature of the equation. These recurrence relations are not explicitly in [8], but after a careful geometric analysis, Biane was able to identify it as discrete Painlevé equations corresponding to an $A_3^{(1)}$ surface in Sakai's classification.

5.6 Semi-classical extensions of Askey–Wilson polynomials

So far all the semi-classical orthogonal polynomials that we encountered were obtained by deformations of the classical weights, i.e., weight functions satisfying a Pearson equation (1.13) with polynomials σ and τ for which deg $\sigma \leq 2$ and deg $\tau = 1$. All the classical orthogonal polynomials satisfy such a Pearson equation, but the differential operator can be replaced by a difference operator, a divided difference operator, a q-difference operator or a divided q-difference operator on special nonuniform lattices. The resulting orthogonal polynomials are given in the Askey table of hypergeometric orthogonal polynomials [101, p. 183] or its q-analogue for basic hypergeometric orthogonal polynomials [101, p. 413]. The most general system of orthogonal polynomials in this Askey scheme are the Askey–Wilson polynomials $p_n(x; a, b, c, d|q)$ and the q-Racah polynomials $R_n(\mu(x); \alpha, \beta, \gamma, \delta|q)$. Both families are related and depend on four parameters. The Askey–Wilson polynomials [101, §14.1] correspond to parameters a, b, c, d which are real or appear in complex conjugate pairs, and when $\max\{|a|, |b|, |c|, |d|\} < 1$ they are orthogonal on $[-1, 1]$ with the weight function

$$w(x) = \frac{1}{\sqrt{1 - x^2}} \left| \frac{(e^{2i\theta}; q)_\infty}{(ae^{i\theta}; q)_\infty (be^{i\theta}; q)_\infty (ce^{i\theta}; q)_\infty (de^{i\theta}; q)_\infty} \right|^2, \qquad 0 < q < 1.$$

If $\max\{|a|, |b|, |c|, |d|\} > 1$ then there is also a discrete part in the orthogonality measure, e.g., for $a > 1$ this will be at the points $x_k = \frac{aq^k + (aq^k)^{-1}}{2}$.

The q-Racah polynomials [101, §14.2] are orthogonal on the quadratic lattice $\mu(x) = q^{-x} + \gamma\delta q^{x+1}$ for $x \in \{0, 1, 2, \ldots, N\}$ when $\alpha q = q^{-N}$, or $\beta\delta q = q^{-N}$, or $\gamma q = q^{-N}$. All other classical orthogonal polynomials in the Askey table can be obtained from the Askey–Wilson polynomials by a limiting procedure. Hence one may argue that there is no need to investigate every particular family of classical orthogonal polynomials, since a study of the Askey–Wilson polynomials would in principle give all information of its limiting cases as well. This is not quite true, since the limiting procedure is not always uniform in all the parameters (including the degree) and hence some properties are lost and new properties may appear for the limiting cases.

   With a similar reasoning, semi-classical weights obtained by deforming a classical weight can be studied from the top level onward, i.e., by deforming the Askey–Wilson weight, and then taking an appropriate limiting procedure. This is precisely what Nicholas Witte set out to do in a long paper [152]. The principal role in that paper is played by a divided difference operator $\mathbb{D}_x$ and an average operator $\mathbb{M}_x$. They are defined by

$$\mathbb{D}_x f(x) = \frac{f(\iota_+(x)) - f(\iota_-(x))}{\iota_+(x) - \iota_-(x)},$$

where $\iota_{\pm}(x)$ are the two y-roots of the quadratic equation

$$Ay^2 + 2Bxy + Cx^2 + 2Dy + 2Ex + F = 0. \tag{5.22}$$

For a fixed y, the two roots of (5.22) define two consecutive points $x(s), x(s + 1)$ on the x-lattice, and conversely for a given x the quadratic (5.22) defines two consecutive points $y(s) = \iota_-(x(s)), y(s + 1) = \iota_+(x(s))$ on the dual lattice. Depending on the value of $B^2 - AC$ and the determinant

$$\det \begin{pmatrix} A & B & D \\ B & C & E \\ D & E & F \end{pmatrix}$$

the lattices are linear (forward difference), q-linear (q-difference), quadratic (divided difference) or q-quadratic (Askey–Wilson operator). For the Askey–Wilson operator we refer to [91, Ch. 16] or [89]. The average operator is given by

$$\mathbb{M}_x f(x) = \frac{f(\iota_+(x)) + f(\iota_-(x))}{2}.$$

The theory of orthogonal polynomials on these special nonuniform lattices of quadratic type was initiated earlier by Magnus [114] and even before that by Askey and Wilson [3] and Nikiforov, Suslov and Uvarov [124].

Witte [152] defines a $\mathbb{D}$-semi-classical weight as a weight w satisfying the Pearson-type equation

$$\mathbb{D}_x w(x) = \frac{2V(x)}{W(x)} \mathbb{M}_x w(x), \qquad (5.23)$$

where V and W are polynomials. For the orthogonality one uses a special integral ($\mathbb{D}$-integral) defined on the x-lattice by

$$\int f(x) \, \mathbb{D}x = \sum_{s \in \mathbb{Z}} (\iota_+(s) - \iota_-(s)) f(x(s)).$$

The orthogonal polynomials satisfy

$$\int p_n(x) \ell_m(x) w(x) \, \mathbb{D}x = 0, \qquad 0 \le m < n,$$

where ℓ_m are polynomial of degree m which form a canonical basis (depending on the lattice) for the polynomials. There is again a three term recurrence relation as in (1.2), and if we define

$$Y_n(x) = \begin{pmatrix} p_n(x) & q_n(x)/w(x) \\ p_{n-1}(x) & q_{n-1}(x)/w(x) \end{pmatrix},$$

where q_n are functions of the second kind

$$q_n(x) = \int \frac{p_n(y)}{x - y} w(y) \, \mathbb{D}y,$$

then

$$Y_{n+1}(x) = B_n(x) Y_n(x), \qquad B_n(x) = \frac{1}{a_{n+1}} \begin{pmatrix} x - b_n & -a_n \\ a_{n+1} & 0 \end{pmatrix}. \qquad (5.24)$$

The Pearson equation (5.23) also implies a $\mathbb{D}$-structure relation of the form

$$W_n(x) \mathbb{D}_x Y_n(x) = A_n(x) \mathbb{M}_x Y_n(x), \qquad (5.25)$$

where

$$A_n(x) = \begin{pmatrix} \Omega_n(x) & -a_n \Theta_n(x) \\ a_n \Theta_{n-1}(x) & -\Omega_n(x) - 2V(x) \end{pmatrix}$$

is a matrix with polynomial coefficients and the degrees of the polynomials W_n, Ω_n and Θ_n are independent of n. The compatibility between (5.24) and (5.25) gives nonlinear recurrence relations for the recurrence coefficients $(a_n, b_n)_{n \in \mathbb{N}}$ or related quantities. The computations however are quite involved.

Witte [152, §7] worked out the simplest extension of the Askey–Wilson weight by taking polynomials V and W satisfying

$$W \pm \Delta y V = z^{\mp 3} \prod_{j=1}^{6} (1 - a_j q^{-1/2} z^{\pm 1}),$$

which has six parameters $a_1, \ldots, a_6$. The four parameters $a_1, \ldots, a_4$ correspond to the four parameters a, b, c, d in the Askey–Wilson weight. For the two remaining parameters the choice $a_5 = \alpha t$ and $a_6 = \alpha/t$ was made, thereby introducing a deformation parameter α and a deformation variable t. Let $(\sigma_1, \ldots, \sigma_6)$ be the elementary symmetric functions of the parameters $q^{-1/2}(a_1, \ldots, a_6)$ and define

$$[s] = q^s \sigma_6 - q^{-s}, \quad \{s\} = q^s \sigma_6 + q^{-s},$$

then the recurrence coefficients a_n and b_n are given by [152, Prop. 7.5]

$$16 \left[n + \frac{1}{2} \right] [n]^2 \left[n - \frac{1}{2} \right] a_n^2 = \sigma_6 w_n^2 + 2(\sigma_1 \sigma_6 + \sigma_5)\{n\} w_n + 2[n]^3 v_n$$

$$+2[n]^2 (4\sigma_6 + 2\sigma_4 + 2\sigma_2 \sigma_6 + 2\sigma_1 \sigma_5 - \{n\}(1 + \sigma_2 + \sigma_4 + \sigma_6) + 2[n]^2) + 4(\sigma_1 \sigma_6 + \sigma_5)^2,$$

and

$$[n+1][n]b_n = -\frac{1}{4} \left\{ n + \frac{1}{2} \right\} w_n - \left[n + \frac{1}{2} \right] [n]\lambda_n - \frac{1}{2}(q^{1/2} + q^{-1/2})(\sigma_1 \sigma_6 + \sigma_5),$$

where λ_n is the zero of the polynomial Θ_n (which is of degree 1) in the matrix A_n. The sequences $(w_n, v_n)_{n \in \mathbb{N}}$ satisfy a coupled system of recurrence relations [152, Prop. 7.4]

$$w_{n+1} - w_n = -4(q^{1/2} - q^{-1/2}) \left[n + \frac{1}{2} \right] \lambda_n,$$

$$v_{n+1} + v_n = -\frac{[n + \frac{1}{2}]}{[n+1][n]} \left(2 \left\{ n + \frac{1}{2} \right\} \lambda_n w_n + 8 \left[n + \frac{1}{2} \right] [n]\lambda_n^2 \right.$$

$$\left. + 4(q^{1/2} + q^{-1/2})(\sigma_1 \sigma_6 + \sigma_5)\lambda_n \right),$$

with initial conditions $w_0 = -2(\sigma_1 + \sigma_5)$ and $v_0 = 1 - \sigma_2 + \sigma_4 - \sigma_6$. This looks like a first order system of equations, but the λ_n still depend on the recurrence coefficients (a_n, b_n), so that one really has nonlinear recurrences.

Witte also gives the t-evolution as a pair of coupled equations, which I am not going to reproduce here because of the complexity of the system, but he claims [152, p. 227] that he *offered unambiguous evidence* that this system is one of the $E_7^{(1)}$ q-Painlevé cases in the Sakai scheme.

6

Special solutions of Painlevé equations

The Painlevé equations are nonlinear differential (and difference) equations and therefore one expects that the solutions are new transcendental functions. This is true for many of the solutions, which in general are meromorphic analytic functions, and some of these transcendental solutions have already been given a name, such as the Ablowitz–Segur and the Hastings–MacLeod solutions of Painlevé II [141] [84] and Boutroux' *tritronquée* solutions of Painlevé I [94]. In this chapter we will see that the Painlevé equations also have rational solutions and solutions in terms of special functions for particular values of the parameters in the equation. We will mostly mention those solutions which are related to orthogonal polynomials.

6.1 Rational solutions

Except for Painlevé I, which has no parameters, there are rational solutions of the Painlevé equations for particular values of the parameters. See the survey papers of Airault [1] and Clarkson [39] [40] for an overview. These rational solutions are in terms of special families of polynomials which can be defined in terms of determinants of more familiar polynomials, such as Hermite and Laguerre polynomials and some other polynomials with simple generating functions.

6.1.1 Painlevé II

The existence of rational solutions of Painlevé II was first observed by Yablonskii [156] and Vorobiev [151].

Theorem 6.1 *The Painlevé II equation* (1.19) *has rational solutions if and* *only if $\alpha = n \in \mathbb{Z}$. The solutions for $\alpha = n$ are given by*

$$w_n(z) = \frac{d}{dz} \log \frac{Q_{n-1}(z)}{Q_n(z)}, \tag{6.1}$$

where Q_n are monic polynomials of degree $n(n+1)/2$ which satisfy the recurrence relation

$$Q_{n+1}(z)Q_{n-1}(z) = zQ_n^2(z) + 4(Q_n'(z))^2 - 4Q_n(z)Q_n''(z), \tag{6.2}$$

with initial values $Q_0(z) = 1$ and $Q_1(z) = z$.

The polynomials $(Q_n)_n$ are nowadays known as *Yablonskii–Vorobiev* polynomials. The fact that the recurrence relation (6.2) gives polynomials is already surprising, since in order to obtain $Q_{n+1}(z)$ one needs to divide by $Q_{n-1}(z)$ and it is not obvious that the right-hand side has this polynomial as a common factor. Kajiwara and Ohta [97] showed that the Yablonskii–Vorobiev polynomials can be represented in terms of a determinant of an easier family of polynomials $(p_n)_{n \in \mathbb{N}}$:

$$\tau_n(z) = \det \begin{pmatrix} p_1(z) & p_3(z) & p_5(z) & \cdots & p_{2n-1}(z) \\ p_1'(z) & p_3'(z) & p_5'(z) & \cdots & p_{2n-1}'(z) \\ \vdots & \vdots & \vdots & \cdots & \vdots \\ p_1^{(n-1)}(z) & p_3^{(n-1)}(z) & p_5^{(n-1)}(z) & \cdots & p_{2n-1}^{(n-1)} \end{pmatrix},$$

where the $(p_n)_n$ have the generating function

$$\sum_{k=0}^{\infty} p_k(z)\lambda^k = \exp(z\lambda - \frac{4}{3}\lambda^3). \tag{6.3}$$

The relation with the Yablonskii–Vorobiev polynomial $Q_n(z)$ is that $\tau_n(z) = c_n Q_n(z)$, with

$$c_n = \prod_{j=1}^{n} (2j+1)^{j-n}.$$

This constant c_n is irrelevant for the rational solution (6.1) since it disappears after differentiation. It is easy to see from the generating function that $p_n'(z) = p_{n-1}(z)$, and using this and transposition gives the equivalent determinant

$$\tau_n(z) = \det \begin{pmatrix} p_n(z) & p_{n+1}(z) & p_{n+2}(z) & \cdots & p_{2n-1}(z) \\ p_{n-2}(z) & p_{n-1}(z) & p_n(z) & \cdots & p_{2n-3}(z) \\ \vdots & \vdots & \vdots & \cdots & \vdots \\ p_{-n+2}(z) & p_{-n+3}(z) & p_{-n+5}(z) & \cdots & p_1(z) \end{pmatrix},$$

with $p_k(z) = 0$ whenever $k < 0$.

> **Exercise 12:** Use the generating function (6.3) to show that the polynomials $(p_n)_n$ satisfy the recurrence relation
>
> $$zp_n(z) = (n+1)p_{n+1}(z) + 4p_{n-2}(z),$$
>
> and the differential equation
>
> $$4p_n'''(z) - zp_n'(z) + np_n(z) = 0.$$

The polynomials $(p_n)_{n \in \mathbb{N}}$ are not orthogonal polynomials but they are *multiple* orthogonal polynomials, closely related to the multiple orthogonal polynomials associated to modified Bessel functions $K_\nu, K_{\nu+1}$ introduced in [150], a fact which was not observed before in the literature.

Property 6.2 *Let $p_{3n}(z) = q_n(z^3)$, $p_{3n+1}(z) = zr_n(z^3)$ and $p_{3n+2}(z) = z^2 s_n(z^3)$. Let $y_{n,m}(x) = y_{n,m}(x; \alpha, \nu)$ be the multiple orthogonal polynomials satisfying*

$$\int_0^\infty y_{n,m}(x) x^\alpha K_\nu(2\sqrt{x}) x^k \, dx = 0, \qquad 0 \le k \le n-1,$$

$$\int_0^\infty y_{n,m}(x) x^\alpha K_{\nu+1}(2\sqrt{x}) x^k \, dx = 0, \qquad 0 \le k \le m-1.$$

Then the polynomials are related by $q_n(36x) = y_{n,n}(x; -2/3, 1/3)$, $r_n(36x) = y_{n,n}(x; 1/3, -2/3)$ and $s_n(36x) = y_{n,n}(x; 2/3, -1/3)$.

Note that the Bessel function $K_{1/3}$ is related to the Airy function. The multiple orthogonality can immediately be expressed in terms of the Airy function as

$$\int_\Gamma p_n(z) \mathrm{Ai}(2^{-2/3}|z|) z^k \, dz = 0, \qquad 0 \le k \le \lceil \tfrac{n}{2} \rceil - 1,$$

$$\int_\Gamma p_n(z) \frac{|z|}{z} \mathrm{Ai}'(z)(2^{-2/3}|z|) z^k \, dz = 0, \qquad 0 \le k \le \lfloor \tfrac{n}{2} \rfloor - 1,$$

where $\Gamma = \Gamma_0 \cup \Gamma_1 \cup \Gamma_2$ and Γ_k is the half line starting at 0 with slope $2\pi k/3$:

$$\Gamma_k = \{z \in \mathbb{C} : \arg(z) = \frac{2\pi k}{3}\}, \qquad k = 0, 1, 2.$$

Unfortunately, this observation does not immediately give more insight in the Yablonskii–Vorobiev polynomials.

A determinant representation for the square of Yablonskii–Vorobiev polynomials was given by Bertola and Bothner [7]:

$$Q_{n-1}^2(z) = d_n \det \begin{pmatrix} \mu_0(z) & \mu_1(z) & \mu_2(z) & \cdots & \mu_{n-1}(z) \\ \mu_1(z) & \mu_2(z) & \mu_3(z) & \cdots & \mu_n(z) \\ \vdots & \vdots & \vdots & \cdots & \vdots \\ \mu_{n-1}(z) & \mu_n(z) & \mu_{n+1}(z) & \cdots & \mu_{2n-2}(z) \end{pmatrix}, \tag{6.4}$$

where

$$d_n = (-1)^{\lfloor \frac{n}{2} \rfloor} 2^{1-n} \prod_{k=1}^{n} \left(\frac{(2k)!}{k!} \right)^2,$$

and $\mu_k(z) = p_k(2^{2/3}z)2^{-2k/3}$, with $(p_n)_n$ the polynomials from the generating function (6.3). The representation (6.4) has the advantage that it involves a Hankel determinant, i.e., a determinant of the form $\det(m_{i+j})_{i,j=1}^n$, see (1.7), and Hankel determinants are closely related to orthogonal polynomials. If one can recognize the $(\mu_k)_k$ in the Hankel matrix as moments of a measure μ in the complex plane, then the monic orthogonal polynomials $(P_n)_n$ with

$$\int P_n(x)P_m(x)\,d\mu(x) = 0, \qquad m \neq n,$$

are given by, see (1.8),

$$P_n(x) = \frac{1}{\Delta_n} \det \begin{pmatrix} \mu_0 & \mu_1 & \mu_2 & \cdots & \mu_n \\ \mu_1 & \mu_2 & \mu_3 & \cdots & \mu_{n+1} \\ \vdots & \vdots & & \cdots & \vdots \\ \mu_{n-1} & \mu_n & \mu_{n+1} & \cdots & \mu_{2n-1} \\ 1 & x & x^2 & \cdots & x^n \end{pmatrix},$$

where Δ_n is the Hankel determinant

$$\Delta_n = \det \begin{pmatrix} \mu_0 & \mu_1 & \mu_2 & \cdots & \mu_{n-1} \\ \mu_1 & \mu_2 & \mu_3 & \cdots & \mu_n \\ \vdots & \vdots & & \cdots & \vdots \\ \mu_{n-1} & \mu_n & \mu_{n+1} & \cdots & \mu_{2n-2} \end{pmatrix}.$$

When μ is a positive measure on the real line with infinite support, then the Hankel matrix $(\mu_{i+j-2})_{i,j=1}^n$ is positive definite and hence $\Delta_n > 0$ for all n, so that all the orthogonal polynomials exist. When μ is not a positive measure or when it is a measure in the complex plane, then some of the Δ_n may be zero and then the corresponding monic orthogonal polynomials do not exist. For the Hankel determinant in (6.4) the moments depend on the variable z and hence the corresponding measure μ also depends on z. We see that $\Delta_n = 0$ if and only if $Q_{n-1}(z) = 0$, hence whenever z is a zero of Q_{n-1}. Therefore the location of the zeros of Q_{n-1} tells us for which z the underlying orthogonal polynomials do not exist. This gives a way to investigate the zeros of Yablonskii–Vorobiev polynomials in terms of existence of the underlying orthogonal polynomials. This was exploited by Bertola and Bothner in [7]. A similar approach using Hankel determinants can also be used for the zeros and poles of rational solutions of Painlevé IV, and we will give more details of this approach in Section 6.1.3. In Figure 6.1 one can see that the zeros of the Yablonskii–Vorobiev polynomials form a nice geometric pattern in the complex plane and (after scaling them by a factor $n^{2/3}$) seem to fill a triangle. This is not really the case. Bertola and Bothner [7] and Buckingham and Miller [22] [23] (see also Miller and Sheng [121]) have obtained precise information about the asymptotic behavior of the zeros of $Q_n(n^{2/3}z)$, i.e., the zeros of the Yablonskii–Vorobiev polynomials after scaling them by a factor $n^{2/3}$.

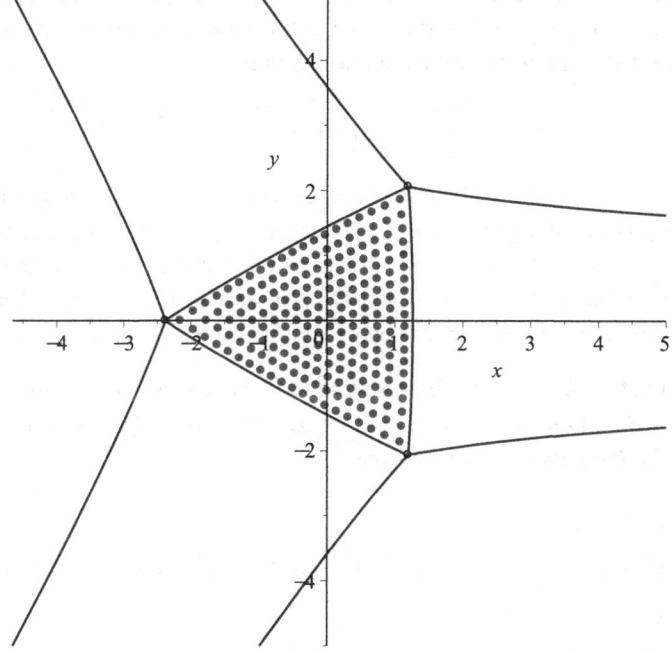

Figure 6.1 The curves given by (6.5) and the zeros of Q_{20}

Theorem 6.3 (Bertola–Bothner, Buckingham–Miller)   *The zeros of $Q_n(n^{2/3}z)$ accumulate on a triangular shaped region Ω with corners $x_k = -3/\sqrt[3]{2}e^{\frac{2\pi i}{3}k}$,* *k = 0, 1, 2, and sides given by*

$$\Re\left(-2\log\frac{1+\sqrt{1+2a^3}}{ia\sqrt{2a}} + \sqrt{1+2a^3}\frac{4a^3-1}{3a^3}\right) = 0, \qquad (6.5)$$

where $a = a(x)$ is a solution of the cubic equation $1 + 2xa^2 - 4a^3 = 0$.

The three corners are precisely the zeros of the discriminant of the cubic equation. The equation (6.5) describes a number of contours emanating from these three corner points, but only those contours that remain bounded and connect the three corners are needed for the boundary of Ω. See Figure 6.1 for these curves and the zeros of the Yablonskii–Vorobiev polynomial Q_{20}.

6.1.2 Painlevé III

We will consider the generic case of P$_{\text{III}}$ in (1.20) with $\gamma\delta \neq 0$ and without loss of generality we put $\gamma = 1$ and $\delta = -1$ (otherwise, a rescaling of y and/or x will be needed). The differential equation then is

$$y'' = \frac{(y')^2}{y} - \frac{y'}{z} + \frac{\alpha y^2 + \beta}{z} + y^3 - \frac{1}{y}. \qquad (6.6)$$

The parameters α and β which give rational solutions were found by Lukashevich [109]. Umemura [148] found a sequence of polynomials (in $1/z$) which give the rational solutions. In the following we use polynomials in the variable z to describe these rational solutions. They are nowadays known as *Umemura polynomials*. The following result can be found in [95] and [36].

Theorem 6.4 *The Painlevé III equation in (6.6) has rational solutions if and only if $\alpha + \beta = 4n$ or $\alpha - \beta = 4n$, for $n \in \mathbb{Z}$. Let S_n be polynomials (of degree $n(n+1)/2$) that satisfy the recurrence*

$$S_{n+1}(z;\mu)S_{n-1}(z;\mu)$$
$$= -z[S_n(z;\mu)S_n''(z;\mu) - (S_n'(z;\mu))^2] - S_n(z;\mu)S_n'(z;\mu) + (z+\mu)S_n^2(z;\mu),$$
$$(6.7)$$

with $S_{-1} = S_0 = 1$, then

$$w_n(z) = 1 + \frac{d}{dz}\log\frac{S_{n-1}(z;\mu-1)}{S_n(z;\mu)} = \frac{S_n(z;\mu-1)S_{n-1}(z;\mu)}{S_n(z;\mu)S_{n-1}(z;\mu-1)} \qquad (6.8)$$

are rational solutions of (6.6) for $\alpha = 2n + 2\mu - 1$ and $\beta = 2n - 2\mu + 1$, and

$$v_n(z) = 1 - \frac{d}{dz}\log\frac{S_{n-1}(z;\mu)}{S_n(z;\mu-1)} = \frac{S_n(z;\mu)S_{n-1}(z;\mu-1)}{S_n(z;\mu-1)S_{n-1}(z;\mu)} \qquad (6.9)$$

are rational solutions for $\alpha = -2n + 2\mu - 1$ and $\beta = -2n - 2\mu + 1$. The rational solutions for $\alpha - \beta = \pm 4n$ are given by $iw_n(iz)$ and $iv_n(iz)$.

Again it is not obvious that the recurrence relation (6.7) gives polynomials. Observe that $v_n(z) = 1/w_n(z)$.

Kajiwara and Masuda [95] obtained an expression involving determinants: let

$$
\tau_n(z) = \det \begin{pmatrix}
L_n^{\mu-n}(-z) & L_{n+1}^{\mu-n-1}(-z) & \cdots & L_{2n-1}^{\mu-2n+1}(-z) \\
L_{n-2}^{\mu-n+2}(-z) & L_{n-1}^{\mu-n+1}(-z) & \cdots & L_{2n-3}^{\mu-2n+3}(-z) \\
\vdots & \vdots & \cdots & \vdots \\
L_{-n+2}^{\mu+n-2}(-z) & L_{-n+3}^{\mu+n-3}(-z) & \cdots & L_1^{\mu-1}(-z)
\end{pmatrix}
\tag{6.10}
$$

where L_n^α are the Laguerre polynomials, then $\tau_n(z) = c_n S_n(z)$ with

$$
c_n = \prod_{j=0}^{n-1}(2j+1)^{j-n}.
$$

This constant c_n is irrelevant for the rational solution. The Laguerre polynomials L_n^α satisfy the differentiation property $(L_n^\alpha(x))' = -L_{n-1}^{\alpha+1}(x)$, hence the polynomials $p_n(x) = L_n^{\mu-n}(-x)$, which appear in the determinant formula, satisfy $p_n'(x) = p_{n-1}(x)$. If we use this, then an equivalent determinant formula is

$$
\tau_n(x) = \det \begin{pmatrix}
L_1^{\mu-1}(-x) & L_3^{\mu-3}(-x) & \cdots & L_{2n-1}^{\mu-2n+1}(-x) \\
[L_1^{\mu-1}(-x)]' & [L_3^{\mu-3}(-x)]' & \cdots & [L_{2n-1}^{\mu-2n+1}(-x)]' \\
\vdots & \vdots & \cdots & \vdots \\
[L_1^{\mu-1}(-x)]^{(n-1)} & [L_3^{\mu-3}(-x)]^{(n-1)} & \cdots & [L_{2n-1}^{\mu-2n+1}(-x)]^{(n-1)}
\end{pmatrix},
\tag{6.11}
$$

which is a Wronskian determinant. The Laguerre polynomials are given by

$$
L_n^\alpha(x) = \sum_{k=0}^{n} \binom{n+\alpha}{n-k} \frac{(-x)^k}{k!},
$$

from which it is easy to find the generating function

$$
\sum_{k=0}^{\infty} p_k(x)\lambda^k = \sum_{k=0}^{\infty} L_k^{\mu-k}(-x)\lambda^k = (1+\lambda)^\mu e^{x\lambda}.
$$

This determinant representation for Umemura polynomials therefore has many similarities with the determinant representation for Yablonskii–Vorobiev polynomials.

The zeros of S_n, or the zeros of τ_n, again have a nice geometric structure, but the structure depends on the parameter μ. If μ is positive and large, then

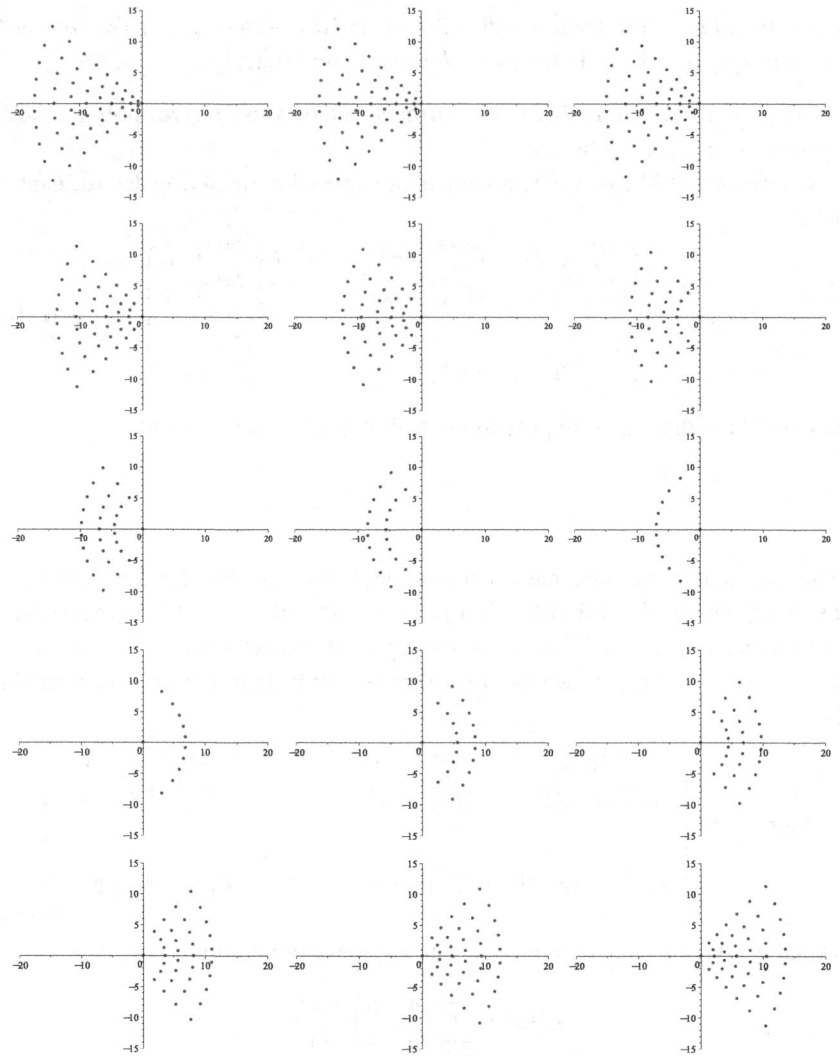

Figure 6.2 Zeros of $S_{10}(z;\mu)$ for $\mu = k$, $9 \geq k \geq -6$

the zeros of $S_n(z,\mu)$ are lying in a triangular shaped region. If $\mu = 0$ then all the zeros are at the origin. If μ is negative and large, then all the zeros are again in a triangular region, but now mirrored along the imaginary axis. For μ an integer k with $-n < k < n$ there are multiple zeros at the origin. In Figure 6.2 the zeros of the Umemura polynomial $S_{10}(z;\mu)$ (or τ_{10}) are plotted for $\mu = 9, 8, 7, 6, 5, 4, 3, 2, 1, -1, -2, -3, -4, -5, -6$. For μ negative the zeros are on the right-hand side of the plane, since $S_n(z; -\mu) = S_n(-z; \mu)$ and for $\mu = 0$ all the zeros are at 0. The asymptotic behavior of these zeros, after scaling, is under investigation by Bothner, Miller and Sheng [19].

6.1.3 Painlevé IV

For the Painlevé IV equation (1.21) the rational solutions appear for the following parameters (see [39] and the references there)

Theorem 6.5 *The fourth Painlevé equation* (1.21) *has rational solutions if and only if*

$$\alpha = m, \quad \beta = -2(2n - m + 1)^2,$$

or

$$\alpha = m, \quad \beta = -2(2n - m + \frac{1}{3})^2,$$

where $m, n \in \mathbb{Z}$.

Okamoto [135] found systems of polynomials with which the rational solutions can be obtained. These families were generalized by Noumi and Yamada [128] who introduced the generalized Hermite polynomials and the generalized Okamoto polynomials. See also [37] for more information about these polynomials, but beware that Clarkson uses a different scaling and different indices and we use his notation in what follows. The rational solutions that are related to the first set of parameters in Theorem 6.5 are

$$\frac{d}{dz} \log \frac{H_{m+1,n}(z)}{H_{m,n}(z)}, \quad \frac{d}{dz} \log \frac{H_{m,n}(z)}{H_{m,n+1}(z)}, \quad -2z + \frac{d}{dz} \log \frac{H_{m,n+1}(z)}{H_{m+1,n}(z)}$$

for (α, β) equal to $(2m+n+1, -2n^2), (-(m+2n+1), -2m^2), (n-m, -2(m+n+1)^2)$ respectively, where $H_{m,n}$ are known as *generalized Hermite polynomials*:

$$H_{m,n}(z) = \det \begin{pmatrix} H_m(z) & H_{m+1}(z) & \cdots & H_{m+n-1}(z) \\ H_{m+1}(z) & H_{m+2}(z) & \cdots & H_{m+n}(z) \\ \vdots & \vdots & \cdots & \vdots \\ H_{m+n-1}(z) & H_{m+n}(z) & \cdots & H_{m+2n-2}(z) \end{pmatrix}, \tag{6.12}$$

with $H_m(z)$ the Hermite polynomials. These generalized Hermite polynomials satisfy a system of recurrence relations

$$2mH_{m+1,n}H_{m-1,n} = H_{m,n}H_{m,n}'' - (H_{m,n}')^2 + 2mH_{m,n}^2,$$
$$2nH_{m,n+1}H_{m,n-1} = -H_{m,n}H_{m,n}'' + (H_{m,n}')^2 + 2nH_{m,n}^2,$$

with initial values $H_{0,0} = H_{1,0} = H_{0,1} = 1$ and $H_{1,1}(z) = 2z$.

The asymptotic distribution of the zeros of $H_{m,n}$ for $m \to \infty$ and n fixed has been investigated by Felder, Hemery and Veselov [63] and also by Masoero and Roffelsen [116]. In this case the order of the determinant defining $H_{m,n}$ remains fixed and one can use the asymptotic behavior of the Hermite polynomials to find the asymptotic distribution of the zeros of the generalized Hermite polynomial $H_{m,n}$ as $m \to \infty$. The zeros are lying on a finite number of curves which is illustrated in Figure 6.3 for $H_{40,5}$.

Theorem 6.6 (Masoero and Roffelsen) *The zeros of $H_{m,n}(\sqrt{2m + nz})$, for $m \to \infty$ and n fixed, accumulate on the curves*

$$y = \frac{2j \log\left(2\sqrt{2m + n}(1 - x^2)^{3/4}\right) - \log F_{n,j}}{2(2m + n)(1 - x^2)^{1/2}}, \qquad (6.13)$$

where $z = x + iy$ and $j \in \{-n + 1, -n + 3, \ldots, n - 3, n - 1\}$ and

$$F_{n,j} = \frac{\Gamma(\frac{1}{2}(n + j + 1))}{\Gamma(\frac{1}{2}(n - j + 1))}.$$

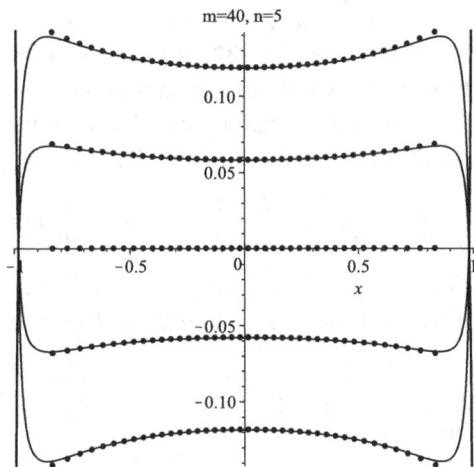

Figure 6.3 Zeros of the generalized Hermite polynomial $H_{40,5}(\sqrt{85}\,z)$ and the curves in (6.13)

The asymptotic distribution of the zeros of generalized Hermite polynomials when both $m, n \to \infty$, with $m/n \to r$, was investigated by Novokshenov and Schelkonogov [130], who used a Riemann–Hilbert problem for Painlevé IV, which reduces to a Riemann–Hilbert problem for orthogonal polynomials.

They claim to have obtained precise asymptotics for the zeros, but their analysis is faulty and they are mixing up the generalized Hermite polynomials with the orthogonal polynomials of their Riemann–Hilbert problem (see also the comment [21, §1.2]). Buckingham gave a correct analysis of the rational solutions associated with generalized Hermite polynomials in [21]. In particular he was able to find an expression for the curves which are the boundary of the square shaped or rectangular shaped region where the zeros of the generalized Hermite polynomials accumulate.

Theorem 6.7 (Buckingham) *The zeros of the generalized Hermite polynomial $H_{m,n}(\sqrt{m}z)$ accumulate in a square or rectangular shaped region with corners given by four solutions of the equation*

$$r^4 x^8 - 24r^2(r^2 + r + 1)x^4 + 32r(2r^3 + 3r^2 - 3r - 2)x^2 - 48(r^2 + r + 1)^2 = 0$$

which are not on the real or imaginary line, and sides given by the curves

$$\Re\left\{ \frac{(1 + r)r^{1/2}xR}{2} - (1 + r)\log\left(2R - \frac{4}{(1 + r)Q} - S\right) \right.$$
$$\left. + (r - 1)\log\left((1 + r)Q^3 + (1 + r)Q^2 R + S\right) + \log(S^2 - 4Q^2) \right\} = 0, \quad (6.14)$$

where $r = m/n \geq 1$. Here $Q = Q(x, r)$ is a solution of the quartic equation

$$3(1 + r)^2 Q^4 + 8(1 + r)r^{1/2}xQ^3 + 4(r - 1 + rx^2)Q^2 - 4 = 0,$$

and $S = S(x, r)$ and $R = R(x, r)$ are given by

$$S = (1 + r)Q^3 + 2r^{1/2}xQ^2,$$

and

$$R = -\frac{\left((1 + r)^2 Q4 + 2(1 + r)QS + 4\right)^{1/2}}{(1 + r)Q}.$$

These curves and the zeros of $H_{20,20}(\sqrt{20}z)$ are plotted in Figure 6.4. The proof of this result is along the same lines as the Bertola–Bothner proof of Theorem 6.3 and uses the observation that (6.12) expresses $H_{m,n}$ as a Hankel determinant. The entries of this Hankel determinant are moments of a (complex measure): one can use the integral representation for the Hermite polynomials [123, Eq. 18.10.10]

$$e^{-z^2} H_n(z) = \frac{(-2i)^n}{\sqrt{\pi}} \int_{-\infty}^{\infty} t^n e^{-t^2} e^{2izt}\, dt$$

Special solutions of Painlevé equations

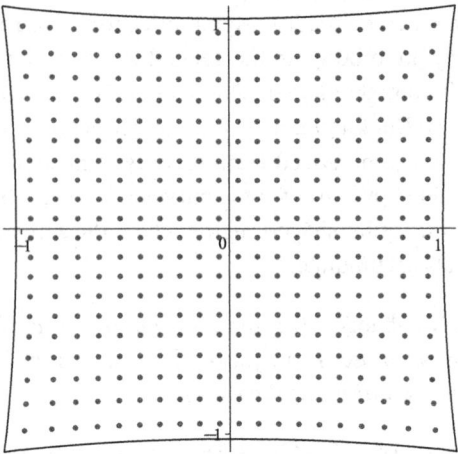

Figure 6.4 Zeros of generalized Hermite polynomials $H_{20,20}$ and the curves in (6.14)

to see that $e^{-z^2} H_{n+m}(z)$ is, up to constants, the n-th moment $\mu_n = \mu_n(z)$ of the complex weight function $w_m(t) = t^m e^{-t^2} e^{2izt}$ on $(-\infty, \infty)$, so that

$$
H_{m,n}(z) = c_{m,n} e^{nz^2} \det \begin{pmatrix}
\mu_0(z) & \mu_1(z) & \mu_2(z) & \cdots & \mu_{n-1}(z) \\
\mu_1(z) & \mu_2(z) & \mu_3(z) & \cdots & \mu_n(z) \\
\mu_2(z) & \mu_3(z) & \mu_{(z)} & \cdots & \mu_{n+1}(z) \\
\vdots & \vdots & \vdots & \cdots & \vdots \\
\mu_{n-1}(z) & \mu_n(z) & \mu_{n+1}(z) & \cdots & \mu_{2n-2}(z)
\end{pmatrix},
$$

with $c_{m,n}$ a constant. The monic orthogonal polynomials $(P_n)_{n\in\mathbb{N}}$ for these moments (or for the weight function w_m) are given by (1.8), where Δ_n is the Hankel determinant. This Hankel determinant is zero if and only if $H_{m,n}(z) = 0$, hence if and only if z is a zero of the generalized Hermite polynomial $H_{m,n}$. So zeros of $H_{m,n}$ correspond to values of z for which the monic orthogonal polynomials P_n do not exist. So the strategy is to determine the region in the complex z-plane where the orthogonal polynomials P_n do not exist. This is achieved by using the Riemann–Hilbert analysis for the orthogonal polynomials P_n. If the Riemann–Hilbert analysis gives a solution for n sufficiently large, then the corresponding z is outside the region where the zeros of $H_{m,n}$ accumulate. The weight function w_m depends on this complex parameter z, and can be written as $w_m(t) = t^m e^{-(t-iz)^2} e^{-z^2}$. It turns out that for large z the zeros of the orthogonal polynomials P_n accumulate on one curve connecting two points in the complex plane, and one can do the whole Riemann–Hilbert analysis without any problem. The analysis essentially reduces to solving a Riemann–Hilbert problem with a constant jump over the curve where the zeros accumulate, and this solution is known as the global parametrix. It can be found by constructing a rational function on a two-sheeted Riemann surface of genus one. So in those cases the orthogonal polynomials P_n exist for n sufficiently large. When z is close to the origin, the zeros of the orthogonal polynomials P_n accumulate on two disjoint curves connecting two pairs of points in the complex plane. The construction of the global parametrix now needs some analysis on a two-sheeted Riemann surface of genus one, and the global parametrix depends on doubly periodic functions (theta functions) and can only be constructed if some condition on the periodicity holds. The region where the zeros of $H_{m,n}$ accumulate corresponds to the genus one case of the Riemann surface for the global parametrix of the Riemann–Hilbert problem for the orthogonal polynomials $(P_n)_{n \in \mathbb{N}}$ for the weight w_m. The curves given by (6.14) give the transition from the genus zero to genus one case of the Riemann–Hilbert problem for the Hankel determinant with moments $\mu_n(z)$.

The rational solutions for the second set of parameters in Theorem 6.5 are

$$-\frac{2}{3}z + \frac{d}{dz}\log\frac{Q_{m+1,n}(z)}{Q_{m,n}(z)}, \quad -\frac{2}{3}z + \frac{d}{dz}\log\frac{Q_{m,n}(z)}{Q_{m,n+1}(z)}, \quad -\frac{2}{3}z + \frac{d}{dz}\log\frac{Q_{m,n+1}(z)}{Q_{m+1,n}(z)}$$

for parameters (α,β) of the form $(-(m+2n), -2(m-\frac{1}{3})^2)$, $(2m+n, -2(n-\frac{1}{3})^2)$, $(n-m, -2(m+n+\frac{1}{3})^2)$ respectively, where $Q_{m,n}$ are known as *generalized Okamoto polynomials*. The original Okamoto polynomials are $Q_n(z) = Q_{n+1,0}(z)$ and $R_n(z) = Q_{n,1}(z)$. Again we use the notation from Clarkson for $Q_{m,n}$ and the relation with the polynomials $Q_{m,n}^{NY}$ of Noumi and Yamada is that $Q_{m,n}(z) = \sqrt{3}^{d_{n,m}} Q_{m+n,n}^{NY}(\sqrt{\frac{2}{3}}x)$, where $d_{n,m} = m^2 + n^2 + mn - m - n$ is the degree of $Q_{m,n}$. They satisfy the recurrence relations

$$Q_{m+1,n}Q_{m-1,n} = \frac{9}{2}[Q_{m,n}Q''_{m,n} - (Q'_{m,n})^2] + [2z^2 + 3(2m+n-1)]Q_{m,n}^2,$$

$$Q_{m,n+1}Q_{m,n-1} = \frac{9}{2}[Q_{m,n}Q''_{m,n} - (Q'_{m,n})^2] + [2z^2 - 3(2n+m-1)]Q_{m,n}^2,$$

with initial values $Q_{0,0} = Q_{1,0} = Q_{0,1} = 1$ and $Q_{1,1}(z) = \sqrt{2}z$. Kajiwara and Ohta [98] and Noumi and Yamada [128] gave a determinant representation of the generalized Okamoto polynomials (see also [37], but be careful since the formula there has to be interpreted correctly). For the original Okamoto polynomials one has

$$Q_n(z) = Q_{n+1,0}(z) = C_n \det \begin{pmatrix} p_{2n}(z) & p_{2n+1}(z) & p_{2n+2}(z) & \cdots & p_{3n-1}(z) \\ p_{2n-3}(z) & p_{2n-2}(z) & p_{2n-1}(z) & \cdots & p_{3n-4}(z) \\ p_{2n-6}(z) & p_{2n-5}(z) & p_{2n-4}(z) & \cdots & p_{3n-7}(z) \\ \vdots & \vdots & \vdots & \cdots & \vdots \\ p_{3-n}(z) & p_{4-n}(z) & p_{5-n}(z) & \cdots & p_2(z) \end{pmatrix},$$

and

$$R_n(z) = Q_{n,1}(z) = D_n \det \begin{pmatrix} p_{2n-1}(z) & p_{2n}(z) & p_{2n+1}(z) & \cdots & p_{3n-2}(z) \\ p_{2n-4}(z) & p_{2n-3}(z) & p_{2n-2}(z) & \cdots & p_{3n-5}(z) \\ p_{2n-7}(z) & p_{2n-6}(z) & p_{2n-5}(z) & \cdots & p_{3n-8}(z) \\ \vdots & \vdots & \vdots & \cdots & \vdots \\ p_{2-n}(z) & p_{3-n}(z) & p_{4-n}(z) & \cdots & p_1(z) \end{pmatrix},$$

where C_n and D_n are constants and $(p_n)_n$ are polynomials with generating function

$$\sum_{k=0}^{\infty} p_k(z)\lambda^k = \exp(2\sqrt{2}z\lambda + 6\lambda^2),$$

and $p_n(z) = 0$ if $n < 0$. The generating function for Hermite polynomials $H_n(x)$ is

$$\sum_{k=0}^{\infty} \frac{H_k(x)}{k!} t^k = \exp(2xt - t^2),$$

hence by properly identifying the variables (x, t) and (z, λ) we see that

$$p_k(z) = \frac{H_k(-iz/\sqrt{3})}{k!}(i\sqrt{6})^k,$$

hence these are Hermite polynomials on the imaginary axis. The generalized Okamoto polynomials have a determinant formula in terms of Schur functions

$$S(i_1, i_2, \ldots, i_k) = \det \begin{pmatrix} p_{i_1}(z) & p_{i_1+1}(z) & p_{i_1+2}(z) & \cdots & p_{i_1+k-1}(z) \\ p_{i_2-1}(z) & p_{i_2}(z) & p_{i_2+1}(z) & \cdots & p_{i_2+k-2}(z) \\ p_{i_3-2}(z) & p_{i_3-1}(z) & p_{i_3}(z) & \cdots & p_{i_3+k-3}(z) \\ \vdots & \vdots & \vdots & \cdots & \vdots \\ p_{i_k-k+1}(z) & p_{i_k-k+2}(z) & p_{i_k-k+3}(z) & \cdots & p_{i_k}(z) \end{pmatrix}$$

$$(6.15)$$

with the polynomials $(p_k)_k$ and with a partition $i_1 \geq i_2 \geq \cdots \geq i_k$ $(k = 2m + n)$ which consists of two parts: $(i_1, \ldots, i_n) = (m+2n-\epsilon, m+2n-\epsilon-2, \ldots, m+2-\epsilon)$ with $\epsilon = 0$ or 1, and $(i_{n+1}, \ldots, i_{n+2m}) = (m, m, m-1, m-1, \ldots, 1, 1)$, which is a double partition, often written as $(m^2, (m-1)^2, \ldots, 1^2)$. The following formulas with a determinant of order $2m + n$ hold for $n, m \in \mathbb{N}$:

$$Q_{-m,-n}(z) = C_{m,n} S(m + 2n, m + 2n - 2, \ldots, m + 2,$$
$$m, m, m - 1, m - 1, m - 2, m - 2, \ldots, 1, 1),$$

and

$$Q_{n,m+1}(z) = D_{m,n} S(m + 2n - 1, m + 2n - 3, \ldots, m + 1,$$
$$m, m, m - 1, m - 1, \ldots, 1, 1),$$

where $C_{m,n}$ and $D_{m,n}$ are constants. Note that the original Okamoto polynomials correspond to $m = 0$, i.e., $Q_{n+1}(z) = Q_{0,-n}(z) = Q_{n+1,0}(z)$ and $R_n(z) = Q_{n,1}(z)$. In Figure 6.5 we have plotted the zeros of Q_{10} and $Q_{8,8}$.

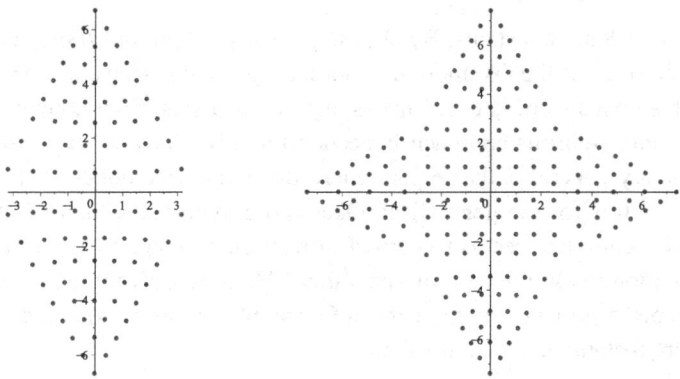

Figure 6.5 Zeros of the Okamoto polynomial Q_{10} (left) and the generalized Okamoto polynomial $Q_{8,8}$ (right)

One easily verifies that $p'_k(z) = 2\sqrt{2} p_{k-1}(z)$, hence all these determinants are Wronskians of the form

$$\det \begin{pmatrix} p_{i_1+k-1}(z) & p_{i_2+k-2}(z) & p_{i_3+k-3}(z) & \cdots & p_{i_k}(z) \\ p'_{i_1+k-1}(z) & p'_{i_2+k-2}(z) & p'_{i_3+k-3}(z) & \cdots & p'_{i_k}(z) \\ p''_{i_1+k-1}(z) & p''_{i_2+k-2}(z) & p''_{i_3+k-3}(z) & \cdots & p''_{i_k}(z) \\ \vdots & \vdots & \vdots & \cdots & \vdots \\ p^{(k-1)}_{i_1+k-1}(z) & p^{(k-1)}_{i_2+k-2}(z) & p^{(k-1)}_{i_3+k-3}(z) & \cdots & p^{(k-1)}_{i_k}(z) \end{pmatrix}$$

with polynomials which are essentially the Hermite polynomials. Such Wronskians recently appeared in the theory of *exceptional orthogonal polynomials*, in particular for *exceptional Hermite polynomials*, see e.g., Gómez-Ullate,

Grandati and Milson [78], Kuijlaars and Milson [104] and Durán [58]. Wronskians containing Laguerre polynomials are used in the construction of *exceptional Laguerre polynomials*, see e.g., Grandati and Quesne [81], Durán [59], Durán and Pérez [61], Bonneux and Kuijlaars [18]. Such Wronskians are more closely related to the Umemura polynomials which we encountered in the previous section and which will appear again in the next section.

Novokshenov and Shchelkonogov [129] have investigated the asymptotic behavior of the Okamoto polynomials. They *approximate* the Painlevé IV equation by the equation

$$V'' = \frac{(V')^2}{2V} + \frac{3}{2}V^3 + 4V^2 + 2AV,$$

which can be integrated once to give the elliptic equation

$$(V')^2 = V^4 + 4V^3 + 4AV^2 + 4BV,$$

where A and B are constants. By choosing appropriate A and B they conclude that the zeros z_k of the Okamoto polynomials $Q_n(\sqrt{n}z)$ behave asymptotically like points on a square lattice. This asymptotic analysis is not convincing and another more rigorous approach is needed that gives precise expressions for the boundary curves of the region where the zeros accumulate. It is a challenge to do this for the generalized Okamoto polynomials: the region where the zeros accumulate seems to consist of a union of a square (or rectangular) shaped region (as in the case of generalized Hermite polynomials) and triangular shaped regions (as in the case of Okamoto polynomials) at the edges of the square/rectangle, see Figure 6.5.

6.1.4 Painlevé V

For the Painlevé V equation (1.22) we will consider only the case $\delta \neq 0$ and without loss of generality we take $\delta = -1/2$. The case $\delta = 0$ can be transformed to Painlevé III (see Theorem 3.9) and has rational solutions related to the rational solutions in Section 6.1.2. So in this section we deal with P$_V$ for a function $w(z)$ of the following form

$$w'' = \left(\frac{1}{2w} + \frac{1}{w-1}\right)(w')^2 - \frac{w'}{z} + \frac{(w-1)^2}{z^2}\left(\alpha w + \frac{\beta}{w}\right) + \frac{\gamma w}{z} - \frac{w(w+1)}{2(w-1)}, \quad (6.16)$$

with three parameters α, β, γ. The rational solutions of (6.16) were characterized by Kitaev, Law and McLeod [100], see also [83], [82, §40], [39, Thm. 5.13]:

Theorem 6.8 *The Painlevé V equation (6.16) has rational solutions if and only if for $m, n \in \mathbb{Z}$*

1. $\alpha = \frac{1}{2}(m \pm 1)^2$ and $\beta = -\frac{n^2}{2}$, with $n > 0$, $m + n$ odd, $\alpha \neq 0$ when $|m| < n$,
2. $\alpha = \frac{n^2}{2}$ and $\beta = -\frac{1}{2}(m \pm 1)^2$, with $n > 0$, $m + n$ odd, $\beta \neq 0$ when $|m| < n$,
3. $\alpha = \frac{a^2}{2}$, $\beta = -\frac{1}{2}(a + n)^2$ and $\gamma = m$, with $m + n$ even and a arbitrary,
4. $\alpha = (b + n)^2$, $\beta = -\frac{b^2}{2}$ and $\gamma = m$, with $m + n$ even and b arbitrary,
5. $\alpha = \frac{1}{8}(2m + 1)^2$, $\beta = -\frac{1}{8}(2n + 1)^2$.

The cases 1 and 2 in this theorem are the special function solutions in terms of the confluent hypergeometric function, which we will describe later in Section 6.2.4, when the confluent hypergeometric functions are Laguerre polynomials. Umemura [148] and Noumi–Yamada [127] found families of polynomials (Umemura polynomials) with which rational solutions of (6.16) can be constructed. Masuda, Ohta and Kajiwara [118] later generalized these and found all the rational solutions from cases 3, 4 and 5 in Theorem 6.8. These generalized Umemura polynomials are given in the next theorem (see [118], [38] and [39, Thm. 5.14]):

Theorem 6.9 *Suppose $U_{m,n}(z; \mu)$ satisfies*

$$U_{m+1,n} U_{m-1,n} = 8z[U_{m,n} U''_{m,n} - (U'_{m,n})^2] + 8U_{m,n} U'_{m,n}$$
$$+ (z + 2\mu - 2 - 6m + 2n)U^2_{m,n},$$
$$U_{m,n+1} U_{m,n-1} = 8z[U_{m,n} U''_{m,n} - (U'_{m,n})^2] + 8U_{m,n} U'_{m,n}$$
$$+ (z - 2\mu - 2 + 2m - 6n)U^2_{m,n},$$

with initial values $U_{-1,-1} = U_{-1,0} = U_{0,-1} = U_{0,0} = 1$. Then $U_{m,n}$ is a polynomial of degree $\frac{1}{2}m(m + 1) + \frac{1}{2}n(n + 1)$ and

$$w(z) = \frac{U_{m,n-1}(z; \mu)U_{m-1,n}(z; \mu)}{U_{m-1,n}(z; \mu - 2)U_{m,n-1}(z; \mu + 2)}$$

is a rational solution of (6.16) for $\alpha = \frac{\mu^2}{8}$, $\beta = -\frac{1}{8}(\mu - 2m + 2n)^2$, $\gamma = -m - n$, and

$$w(z) = \frac{U_{m,n-1}(z; \mu + 1)U_{m,n+1}(z; \mu - 1)}{U_{m-1,n}(z; \mu - 1)U_{m+1,n}(z; \mu + 1)}$$

is a solution of (6.16) for $\alpha = \frac{1}{8}(2m + 1)^2$, $\beta = -\frac{1}{8}(2n + 1)^2$, $\gamma = m - n - \mu$.

The original *Umemura polynomials* are $U_n(z; \mu) = U_{0,n}(z; \mu)$ and $U_{-n}(z; \mu) = U_{n-1,0}(-z; \mu - 1)$ for $n \geq 0$. Again it is quite surprising that the recurrence relations result in polynomials rather than rational functions.

Noumi and Yamada [127] gave a determinantal representation as a Schur function (6.15) for the partition $(n, n - 1, n - 2, \ldots, 2, 1)$ with polynomials

$(p_k^{(r)})_{k\in\mathbb{N}}$ having a generating function

$$\sum_{k=0}^{\infty} p_k^{(r)}(x)\lambda^k = (1-\lambda)^{-r} \exp\left(-\frac{x}{2}\frac{\lambda}{1-\lambda}\right),$$

with $r = \mu + n$. The generating function for Laguerre polynomials is [123, Eq. 18.12.13]

$$\sum_{n=0}^{\infty} L_n^{(\alpha)}(x)z^n = (1-z)^{-\alpha-1} \exp\left(x\frac{z}{z-1}\right), \tag{6.17}$$

hence we see that $p_k^{(r)}(x) = L_k^{(r-1)}(x/2)$. One then has the $n \times n$ determinant

$$U_{0,n}(x;\mu) = C_n \det \begin{pmatrix} p_n^{(r)}(x) & p_{n+1}^{(r)}(x) & \cdots & p_{2n-1}^{(r)}(x) \\ p_{n-2}^{(r)}(x) & p_{n-1}^{(r)}(x) & \cdots & p_{2n-3}^{(r)}(x) \\ p_{n-4}^{(r)}(x) & p_{n-3}^{(r)}(x) & \cdots & p_{2n-5}^{(r)}(x) \\ \vdots & \vdots & \cdots & \vdots \\ p_{2-n}^{(r)}(x) & p_{3-n}^{(r)}(x) & \cdots & p_1^{(r)}(x) \end{pmatrix}, \qquad r = \mu + n, \tag{6.18}$$

with C_n a constant that makes this a monic polynomial (of degree $\frac{n(n+1)}{2}$). One also has the $m \times m$ determinant

$$U_{m,0}(x;\mu) = D_m \det \begin{pmatrix} q_m^{(r)}(x) & q_{m+1}^{(r)}(x) & \cdots & q_{2m-1}^{(r)}(x) \\ q_{m-2}^{(r)}(x) & q_{m-1}^{(r)}(x) & \cdots & q_{2m-3}^{(r)}(x) \\ q_{m-4}^{(r)}(x) & q_{m-3}^{(r)}(x) & \cdots & q_{2m-5}^{(r)}(x) \\ \vdots & \vdots & \cdots & \vdots \\ q_{2-m}^{(r)}(x) & q_{3-m}^{(r)}(x) & \cdots & q_1^{(r)}(x) \end{pmatrix}, \qquad r = \mu - m, \tag{6.19}$$

with D_m a normalizing constant and $q_k^{(r)}(x) = p_k^{(r)}(-x)$, which is also a Schur function for the partition $(m, m-1, \ldots, 2, 1)$ but for the polynomials $(q_k^{(r)})_{k\in\mathbb{N}}$. The generalized Umemura polynomials are a combination of both determinants. For $U_{m,n}(x;\mu)$ one has the $(m+n) \times (m+n)$ determinant

$$C_{m,n} \det \left(\begin{array}{ccccc|ccccc} q_1^{(r)} & q_0^{(r)} & \cdots & q_{-m+2}^{(r)} & q_{-m+1}^{(r)} & q_{-m}^{(r)} & \cdots & q_{-m-n+2}^{(r)} \\ q_3^{(r)} & q_2^{(r)} & \cdots & q_{-m+4}^{(r)} & q_{-m+3}^{(r)} & q_{-m+2}^{(r)} & \cdots & q_{-m-n+4}^{(r)} \\ \vdots & \vdots & \cdots & \vdots & \vdots & \vdots & \cdots & \vdots \\ q_{2m-1}^{(r)} & q_{2m-2}^{(r)} & \cdots & q_m^{(r)} & q_{m-1}^{(r)} & q_{m-2}^{(r)} & \cdots & q_{m-n}^{(r)} \\ \hline p_{n-m}^{(r)} & p_{n-m+1}^{(r)} & \cdots & p_{n-1}^{(r)} & p_n^{(r)} & p_{n+1}^{(r)} & \cdots & p_{2n-1}^{(r)} \\ p_{n-m-2}^{(r)} & p_{n-m-1}^{(r)} & \cdots & p_{n-3}^{(r)} & p_{n-2}^{(r)} & p_{n-1}^{(r)} & \cdots & p_{2n-3}^{(r)} \\ \vdots & \vdots & \cdots & \vdots & \vdots & \vdots & \cdots & \vdots \\ p_{-n-m+2}^{(r)} & p_{-n-m+3}^{(r)} & \cdots & p_{-n+1}^{(r)} & p_{-n+2}^{(r)} & p_{-n+3}^{(r)} & \cdots & p_1^{(r)} \end{array}\right) \tag{6.20}$$

with $r = \mu - m + n$ and $C_{m,n}$ a normalizing constant. Observe that the $m \times m$ corner in the upper left resembles (6.19) and the $n \times n$ corner in the lower right corresponds to (6.18). Note the combined use of the Laguerre polynomials $L_k^{(r-1)}(x/2)$ and $L_k^{(r-1)}(-x/2)$, which is a feature that also happens with exceptional Laguerre polynomials, see [18].

Exercise 13: Show that

$$p_k^{(r)}(x) - p_{k-1}^{(r)}(x) = p_k^{(r-1)}(x), \quad q_k^{(r)}(x) - q_{k-1}^{(r)}(x) = q_k^{(r-1)}(x). \qquad (6.21)$$

Let

$$D = \begin{vmatrix}
-q_1^{(r+1)} & q_1^{(r)} & q_0^{(r)} & \cdots & q_{-m-n+3}^{(r)} & q_{-m-n+2}^{(r)} \\
-q_3^{(r+1)} & q_3^{(r)} & q_2^{(r)} & \cdots & q_{-m-n+5}^{(r)} & q_{-m-n+4}^{(r)} \\
\vdots & \vdots & \vdots & \cdots & \vdots & \vdots \\
-q_{2m-1}^{(r+1)} & q_{2m-1}^{(r)} & q_{2m-2}^{(r)} & \cdots & q_{m-n+1}^{(r)} & q_{m-n}^{(r)} \\
p_{n-m+1}^{(r+1)} & p_{n-m+2}^{(r)} & p_{n-m+3}^{(r)} & \cdots & p_{2n}^{(r)} & p_{2n+1}^{(r)} \\
p_{n-m-1}^{(r+1)} & p_{n-m}^{(r)} & p_{n-m+1}^{(r)} & \cdots & p_{2n-2}^{(r)} & p_{2n-1}^{(r)} \\
\vdots & \vdots & \vdots & \cdots & \vdots & \vdots \\
p_{-n-m+1}^{(r+1)} & p_{-n-m+2}^{(r)} & p_{-n-m+3}^{(r)} & \cdots & p_0^{(r)} & p_1^{(r)}
\end{vmatrix},$$

and denote by $D_{[i,j]}^{[k,\ell]}$ the determinant obtained by deleting rows i, j and columns k, ℓ in D, and similarly $D_{[i]}^{[k]}$ the determinant obtained by deleting row i and column k from D. Let $R_{m,n}(\mu) = U_{m,n}(z; \mu)/C_{m,n}$, with $U_{m,n}$ and $C_{m,n}$ given in (6.20). Show that

$$D = (-1)^m R_{m,n+1}(\mu + 1), \quad D_{[m,m+1]}^{[1,m+n+1]} = R_{m-1,n-1}(\mu)$$

$$D_{[m]}^{[1]} = R_{m-1,n+1}(\mu), \quad D_{[m+1]}^{[1]} = R_{m,n}(\mu),$$

$$D_{[m]}^{[m+n+1]} = (-1)^{m-1} R_{m-1,n}(\mu + 1), \quad D_{[m+1]}^{[m+n+1]} = (-1)^m R_{m,n-1}^{(r+1)}.$$

These generalized Umemura polynomials depend on a parameter μ and the zeros of $U_{m,n}(z; \mu)$ are therefore depending on μ. When μ is negative and large, the zeros are in two well-separated triangular shaped regions, symmetric with respect to the real axis, with $\frac{1}{2}m(m + 1)$ zeros in one triangle (on the right) and $\frac{1}{2}n(n+1)$ in the other triangle (on the left). As μ increases, the triangles move to the middle, and when $\mu = -2n+1, -2n+3, \ldots, -3, -1, 1, 3, \ldots, 2m-3, 2m-1$ zeros will be coalescing at the origin. Once all the zeros have reached the origin, the triangles start to reappear until for μ positive and large there are again two triangular shaped regions, but the two triangles have interchanged

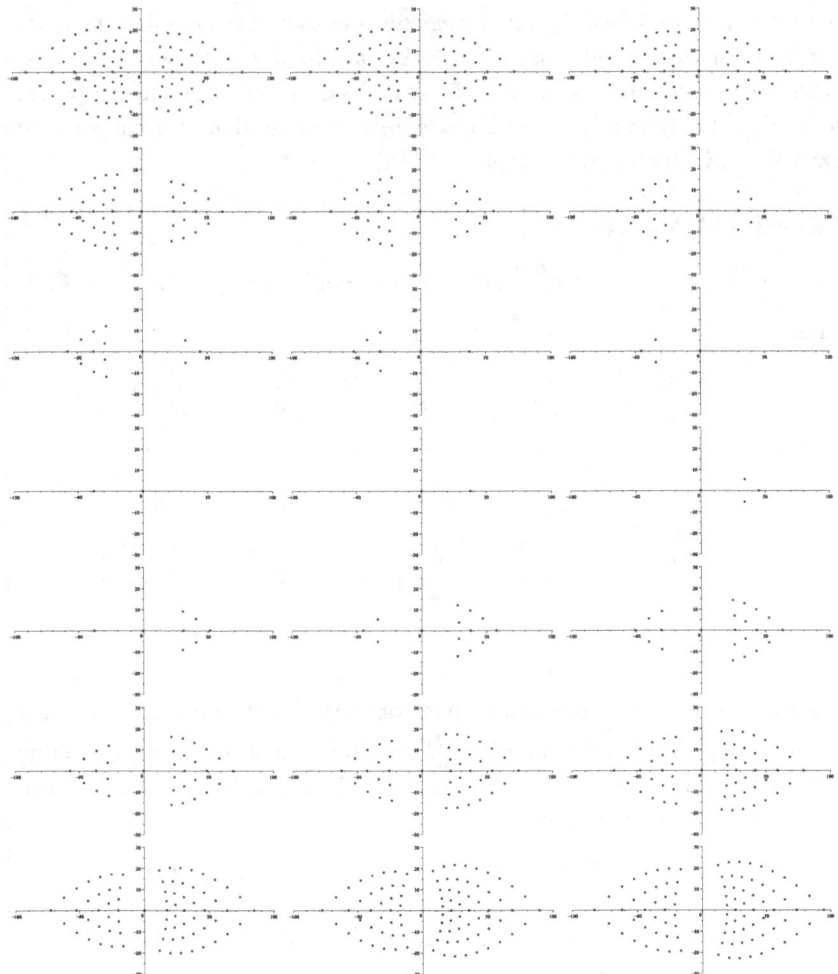

Figure 6.6 Zeros of $U_{8,10}(z;\mu)$ for $\mu = 2k + 1$, $(-11 \le k \le 9)$

their position. See Figure 6.6 for the zeros of $U_{8,10}(z;\mu)$ with odd integer values of μ and [38] for more pictures.

6.1.5 Painlevé VI

Rational solutions of Painlevé VI are characterized in the following result of Mazzocco [120], see also [39, Thm. 5.15]:

Theorem 6.10 *The Painlevé VI equation in* (1.23) *has rational solutions, if* *and only if*

$$a + b + c + d = 2n + 1,$$

with $n \in \mathbb{Z}$, $a = \pm \sqrt{2\alpha}$, $b = \pm \sqrt{-2\beta}$, $c = \pm \sqrt{2\gamma}$ and $d = \pm \sqrt{1 - 2\delta}$, where the $\pm$ signs can be taken independently and at least one of a, b, c, d is an integer.

These rational solutions correspond to the special function solutions in terms of the hypergeometric function (see Section 6.2.5) in the case where the hypergeometric function reduces to a Jacobi polynomial. The rational solutions are in terms of determinants involving the Jacobi polynomial. Very little is known about the location of the zeros of such determinants, but such determinants also appear as exceptional Jacobi polynomials, see, e.g., [56], [60], [79], [110], [140].

6.2 Special function solutions

The Painlevé equations II–VI have special solutions in terms of classical special functions. Painlevé I does not have any parameters and has no special solution of this form. Each of the special function solutions are generated by a solution of a Riccati equation of the form

$$y'(x) = p_2(x)y^2 + p_1(x)y + p_0(x),$$

with rational functions p_0, p_1, p_2. This solution will be a solution of a Painlevé equation for a particular value of the parameter. Bäcklund transformations then will give a whole family of special function solutions generated by this seed solution. Bäcklund transformations transform a solution of a Painlevé equation to another solution of the same Painlevé equation but with different parameter values. The special function solutions usually depend on one parameter (the constant of integration of the Riccati equation) and the special functions often appear in determinants. We will show how these special solutions are related to orthogonal polynomials. For this section we heavily relied on Clarkson's survey [39, §7] which gives many references to papers dealing with these special function solutions.

6.2.1 Painlevé II

The special function solutions for P_{II} are in terms of Airy functions Ai and Bi, which are two linearly independent solutions of the Airy equation

$$w''(x) = xw(x),$$

see [123, §9]. We start from the Riccati equation

$$y'(x) = y^2 + \frac{1}{2}x, \tag{6.22}$$

which after differentiation gives

$$y''(x) = 2y^3 + xy + \frac{1}{2}.$$

This is P_{II} in (1.19) with $\alpha = \frac{1}{2}$. One solves the Riccati equation in the usual way by putting $y(x) = -\varphi'(x)/\varphi(x)$, which gives the second order linear differential equation

$$\varphi'' + \frac{1}{2}x\varphi = 0,$$

and this is the Airy equation, up to a change of the variable x. The general solution is

$$\varphi(x) = c_1 \operatorname{Ai}(-2^{-1/3}x) + c_2 \operatorname{Bi}(-2^{-2/3}x), \tag{6.23}$$

so that the seed solution is

$$y_0(x) = -\frac{\varphi'(x)}{\varphi(x)} = 2^{-1/3}\frac{c_1 \operatorname{Ai}'(-2^{-1/3}x) + c_2 \operatorname{Bi}'(-2^{-2/3}x)}{c_1 \operatorname{Ai}(-2^{-1/3}x) + c_2 \operatorname{Bi}(-2^{-2/3}x)}.$$

Even though this contains two parameters c_1, c_2, it only depends on the ratio c_1/c_2. The Bäcklund transformations

$$y_{n+1} = -y_n + \frac{2(n+1)}{y_n^2 + y_n' - x},$$

$$y_{n-1} = -y_n + \frac{2n}{y_n^2 - y_n' - x}$$

now allow to find special solutions y_n for the parameter $n + \frac{1}{2}$ for every $n \in \mathbb{Z}$ starting from the seed solution y_0. These solutions are

$$y(x; n + \frac{1}{2}) = \frac{\tau_n'}{\tau_n} - \frac{\tau_{n+1}'}{\tau_{n+1}}, \tag{6.24}$$

where $\tau_n = \tau_n(x)$ is the Wronskian

$$\tau_n = \det\begin{pmatrix} \varphi & \varphi' & \varphi'' & \cdots & \varphi^{(n-1)} \\ \varphi' & \varphi'' & \varphi''' & \cdots & \varphi^n \\ \varphi'' & \varphi''' & \varphi^{(4)} & \cdots & \varphi^{(n+1)} \\ \vdots & \vdots & \vdots & \cdots & \vdots \\ \varphi^{(n-1)} & \varphi^{(n)} & \varphi^{(n+1)} & \cdots & \varphi^{(2n-2)} \end{pmatrix},$$

with φ given in (6.23).

Exercise 14: Show that the functions τ_n satisfy the differential-difference equation

$$\tau_{n+1}\tau_{n-1} = \tau_n\tau_n'' - (\tau_n')^2,$$

with initial values $\tau_0 = 1$ and $\tau_1 = \varphi$. This can be used to compute τ_n recursively.

These special solutions appear when one deals with orthogonal polynomials in the complex plane with the weight function

$$w(z) = e^{-\frac{1}{3}z^3 + tz}.$$

One needs to integrate over a curve such that the weight vanishes along that curve, for instance one can use the lines $\Gamma = (e^{2\pi i/3}\infty, 0] \cup [0, \infty)$. Such polynomials were considered in [9] and [50], and related multiple orthogonal polynomials in [66]. The recurrence coefficients are given by

$$a_n^2 = \hat{a}_n e^{-i\pi/3}, \quad b_n = \hat{b}_n e^{i\pi/3},$$

where $\hat{a}_n$ and $\hat{b}_n$ satisfy

$$\begin{cases} \hat{a}_n + \hat{a}_{n+1} = \hat{b}_n - t, \\ \hat{a}_n(\hat{b}_n + \hat{b}_{n-1}) = n. \end{cases}$$

This system of nonlinear equations is known as alternative d-P$_{\text{I}}$. The solution with initial values $\hat{a}_0(t) = 0$ and $\hat{b}_0(t) = -\text{Ai}'(t)/\text{Ai}(t)$ then gives the recurrence coefficients for these orthogonal polynomials. It turns out to be the unique solution with $\hat{a}_0(t) = 0$ for which $\hat{a}_{n+1} > 0$ and $\hat{b}_n > 0$ for all $n \geq 0$, see [44, Thm. 1.1]. The recurrence coefficient $b_n(t)$ is a solution of the differential equation

$$w'' = 2w^3 - 2tw - 2n + 1,$$

and this is related to Painlevé II in (1.19) with $\alpha = n + \frac{1}{2}$ by the change of variables $y(x) = -2^{-1/3}w(t)$ and $x = -2^{1/3}t$. For these orthogonal polynomials one thus has the special solution with $c_2 = 0$ in (6.23), i.e., with only the Airy function Ai. Note that the moments of this exponential weight function are

$$m_n = \frac{1}{2\pi i} \int_\Gamma z^n e^{-\frac{1}{3}z^3 + tz}\, dz,$$

with (see [123, §9, Eq. 9.5.1])

$$m_0 = \frac{1}{2\pi i}\int_\Gamma e^{-\frac{1}{3}z^3 + tz}\, dz = \text{Ai}(t), \qquad m_n = \frac{d^n}{dt^n}m_0.$$

Hence the Hankel determinant (1.7) containing the moments corresponds to

τ_{n+1} and the special solution (6.24) corresponds to the formula (1.11) for b_n in terms of Hankel determinants, see [44, §5].

6.2.2 Painlevé III

The special function solutions for P_{III} are in terms of the Bessel functions J_ν and Y_ν or modified Bessel functions I_ν and K_ν [123, §10]. Taking $\gamma = 1$ and $\delta = -1$ in (1.20), the special function solutions appear when $\alpha + \beta = 4n + 2$ or $\alpha - \beta = 4n + 2$, with $n \in \mathbb{Z}$. Every solution of the Riccati equation

$$xy' = xy^2 + (\alpha - 1)y + x \qquad (6.25)$$

will be a solution of P_{III} when $\alpha + \beta = 2$. To solve the Riccati equation, one puts $y(x) = -\varphi'/\varphi$, giving the second order linear differential equation

$$x\varphi'' - (\alpha - 1)\varphi' + x\varphi = 0,$$

with general solution

$$\varphi(x) = c_1 x^\nu J_\nu(x) + c_2 x^\nu Y_\nu(x), \qquad (6.26)$$

with $\nu = \alpha/2$ and c_1, c_2 constants. This gives the seed solution

$$y_0(x) = -\frac{(c_1 x^\nu J_\nu(x) + c_2 x^\nu Y_\nu(x))'}{c_1 x^\nu J_\nu(x) + c_2 x^\nu Y_\nu(x)}.$$

In a similar way every solution of the Riccati equation

$$xy' = xy^2 + (\alpha - 1)y - x \qquad (6.27)$$

is a solution of P_{III} when $\alpha - \beta = 2$. One now gets the second order linear differential equation

$$x\varphi'' - (\alpha - 1)\varphi' - x\varphi = 0,$$

with general solution

$$\varphi(x) = c_1 x^\nu I_\nu(x) + c_2 x^\nu K_\nu(x), \qquad (6.28)$$

with $\nu = \alpha/2$. The seed solution is now

$$\hat{y}_0(x) = -\frac{(c_1 x^\nu I_\nu(x) + c_2 x^\nu K_\nu(x))'}{c_1 x^\nu I_\nu(x) + c_2 x^\nu K_\nu(x)}.$$

The Bäcklund transformations in [123, §32.7(iii)] then give a family of solutions $(y_n)_{n\in\mathbb{Z}}$ in terms of J_ν and Y_ν with $\alpha + \beta = 4n + 2$, and a family $(\hat{y}_n)_{n\in\mathbb{Z}}$ in terms of I_ν and K_ν) for $\alpha - \beta = 4n+2$. These solutions in terms of a determinant

$$\tau_n = \det \begin{pmatrix} \varphi & \delta\varphi & \delta^2\varphi & \cdots & \delta^{n-1}\varphi \\ \delta\varphi & \delta^2\varphi & \delta^3\varphi & \cdots & \delta^n\varphi \\ \delta^2\varphi & \delta^3\varphi & \delta^4\varphi & \cdots & \delta^{n+1}\varphi \\ \vdots & \vdots & \vdots & \ddots & \vdots \\ \delta^{n-1}\varphi & \delta^n\varphi & \delta^{n+1}\varphi & \cdots & \delta^{2n-2}\varphi \end{pmatrix},$$

with $\delta = x\frac{d}{dx}$ and φ given by (6.26) or (6.28), see [136], [71].

There are a few semi-classical orthogonal polynomials for which the recurrence coefficients are precisely these special function solutions of Painlevé III. The weight function

$$w(x;t) = (1-x^2)^{\nu-\frac{1}{2}}e^{xt}, \qquad x \in [-1,1],$$

is the special case $\alpha = \beta = \nu - \frac{1}{2}$ of the semi-classical weight considered in Section 5.2, but with t replaced by $-t$. The integral

$$I_\nu(t) = \frac{(t/2)^\nu}{\sqrt{\pi}\Gamma(\nu + \frac{1}{2})} \int_{-1}^{1} (1-x^2)^{\nu-\frac{1}{2}}e^{xt}\,dx$$

[123, Eq. 10.32.2] shows that the moment m_0 is $t^{-\nu}I_\nu(t)$ (up to a multiplicative constant), and the moments m_k are

$$\frac{d^k}{dt^k}t^{-\nu}I_\nu(t),$$

up to the same multiplicative constant.

Another semi-classical weight that fits into this family of special function solutions is the weight

$$w(x;t) = x^\alpha e^{-x}e^{-s/x}, \qquad x \in [0,\infty),$$

which we encountered in Section 4.4. The integral representation

$$K_\nu(z) = \frac{1}{2}\left(\frac{z}{2}\right)^\nu \int_0^\infty \exp\left(-t - \frac{z^2}{4t}\right)\frac{dt}{t^{\nu+1}}$$

[123, Eq. 10.32.10] shows that the moments are in terms of the modified Bessel function K_ν.

A third case is the (discrete) generalized Charlier weight from Section 3.2.1

$$w_k = \frac{a^k}{(\beta)_k k!}, \qquad k \in \{0, 1, 2, \ldots\}.$$

Here we use the power series

$$I_\nu(z) = \left(\frac{z}{2}\right)^\nu \sum_{k=0}^{\infty} \frac{(z^2/4)^k}{k!\Gamma(\nu + 1 + k)},$$

to see that $m_0 = a^{-\frac{\beta-1}{2}} \Gamma(\beta) I_{\beta-1}(2\sqrt{a})$. All the other moments can be obtained from this by differentiation. So the Hankel determinant is a determinant containing derivatives of $I_{\beta-1}$ and the recurrence coefficients are expressed in terms of these Hankel determinants. So it is not a surprise that this generalized Charlier weight has recurrence coefficients that satisfy Painlevé III and the solution that we need is the special function solution that involves only the modified Bessel functions I_ν. See also Clarkson [41, §4.2].

6.2.3 Painlevé IV

Special function solutions for P_{IV} exist when $\beta = -2(2n+1 \pm \alpha)^2$ or $\beta = -2n^2$ and they are in terms of parabolic cylinder functions [123, §12]. For $\beta = -2(2n+1+\alpha)$ one used the Riccati equation

$$y' = y^2 + 2xy - 2(1+\alpha) \tag{6.29}$$

and one can verify that each solution of (6.29) is also a solution of P_{IV} in (1.21) when $\beta = -2(1+\alpha)^2$. In a similar way one uses the Riccati equation

$$y' = -y^2 - 2xy - 2(1-\alpha) \tag{6.30}$$

and each solution will also be a solution of (1.21) with $\beta = -2(1-\alpha)^2$.

Exercise 15: Show that each solution of the Riccati equation (6.29) is also a solution of the Painlevé IV equation (1.21) when $\beta = -2(1+\alpha)^2$.

The special function solutions for P_{IV} are now obtained from these seed solutions by using the Bäcklund transformations in [123, §32.7(iv)]. To find the seed solution for $\beta = -2(1+\alpha)^2$, we solve the Riccati equation (6.29) and put $y = -\varphi'/\varphi$, to find that φ satisfies the second order linear differential equation

$$\varphi'' - 2x\varphi' - 2(1+\alpha)\varphi = 0.$$

The general solution of this differential equation is

$$\varphi(x) = \left(c_1 U(\alpha + \frac{1}{2}, \sqrt{2}x) + c_2 V(\alpha + \frac{1}{2}, \sqrt{2}x)\right) e^{x^2/2}, \tag{6.31}$$

where $U(a,z)$ and $V(a,z)$ are two linearly independent solutions of the parabolic cylinder equation

$$w''(z) - \left(\frac{z^2}{4} + a\right) w(z) = 0.$$

For the solution of (6.30) we set $y = \varphi'/\varphi$, to find that φ needs to satisfy

$$\varphi'' + 2x\varphi' + 2(1 - \alpha)\varphi = 0,$$

for which the general solution is

$$\varphi(x) = \left(c_1 U(\alpha - \tfrac{1}{2}, \sqrt{2}x) + c_2 V(\alpha - \tfrac{1}{2}, \sqrt{2}x) \right) e^{-x^2/2}. \qquad (6.32)$$

The special function solutions for $\beta = -2(2n + 1 \pm \alpha)^2$ can again be written in terms of a Wronskian determinant involving the function φ in (6.31) or (6.32).

For the parabolic cylinder function $U(a, z)$ one has the integral representation

$$U(a, z) = \frac{e^{-z^2/4}}{\Gamma(\tfrac{1}{2} + a)} \int_0^\infty x^{a - \frac{1}{2}} e^{-\frac{x^2}{2} - zx} \, dx.$$

The Freud weight from Section 2.1

$$w(x) = e^{-x^4 + tx^2}, \qquad x \in \mathbb{R},$$

has even moments in terms of $U(k, -t)$, and if we consider this weight on the cross (as in Section 2.6), then also the functions $V(k, -t)$ come into play. The recurrence coefficients are thus a special function solution of Painlevé IV, as was found in Section 2.5. This case was also considered in [43].

Another semi-classical weight is the modified Laguerre weight

$$w(x) = x^\alpha e^{-x^2 + tx}, \qquad x \in [0, \infty),$$

which we studied in Section 5.1 and in [42]. The moments are in terms of $U(a, z)$, since the Freud weight is just the symmetric version of this weight with a quadratic change of variable. So the recurrence coefficients are the special function solution of Painlevé IV.

6.2.4 Painlevé V

As usual we will consider only the case when $\delta \neq 0$ and without loss of generality we take $\delta = -\tfrac{1}{2}$. Then P_V has special function solutions if and only if

$$a + b \pm \gamma = 2n + 1 \quad \text{or} \quad (a - n)(b - n) = 0,$$

where $n \in \mathbb{Z}$, $a = \pm \sqrt{2\alpha}$ and $b = \pm \sqrt{-2\beta}$, where the $\pm$ can be taken independently [39, Thm. 7.7], [82, §40]. The special functions are the Kummer functions $M(a, b, x)$ and $U(a, b, x)$, which are given in terms of the confluent hypergeometric function as $M(a, b, z) = {}_1F_1(a; b; z)$ and

$$U(a, b, z) = \frac{\Gamma(1-b)}{\Gamma(a-b+1)} {}_1F_1(a; b; z) + \frac{\Gamma(b-1)}{\Gamma(a)} {}_1F_1(a-b+1; 2-b; z),$$

see [123, §13]. For the case $a+b\pm\gamma = 2n+1$ we start from the Riccati equation

$$xy' = ay^2 + (b - a + x)y - b, \tag{6.33}$$

which after differentiation gives

$$xy'' + y' = 2ay' + (b - a + x)y' + y.$$

Comparison with P_V in (1.22) (with $\delta = -\frac{1}{2}$) shows that a solution of (6.33) is also a solution of Painlevé V when $a + b + \gamma = 1$. To solve the Riccati equation, we put $y = -\frac{x\varphi'}{\varphi}$, to find that φ satisfies the second order linear differential equation

$$x^2\varphi'' + x(1 - b + a - x)\varphi' - ab\varphi = 0.$$

The general solution of this differential equation is

$$\varphi(x) = x^b(c_1 M(b, a + b + 1, x) + c_2 U(b, a + b + 1, x)), \tag{6.34}$$

where the Kummer functions $M(a, b, x)$ and $U(a, b, x)$ are two independent solutions of Kummer's equation

$$xw'' + (b - x)w' - aw = 0. \tag{6.35}$$

Special function solutions of Painlevé V for $a+b+\gamma = 2n+1$ are then obtained from this seed solution by using the Bäcklund transformations [123, §32.7(v)]. These solutions can also be expressed in terms of Wronskian determinants for the function φ in (6.34), see [134], [71], [117]. For the special solution with $a + b - \gamma = 2n + 1$ one uses a seed solution which is a solution of the Riccati equation

$$xy' = ay^2 + (b - a - x)y - b \tag{6.36}$$

and this is a solution of P_V when $a + b - \gamma = 1$. The solution of the Riccati equation is $y' = -\frac{x\varphi'}{a\varphi}$, with

$$\varphi(x) = x^b e^{-x}(c_1 M(a + 1, a + b + 1, x) + c_2 U(a + 1; a + b + 1, x)), \tag{6.37}$$

and via the Bäcklund transformations one then finds the special function solutions for $a + b - \gamma = 2n + 1$.

Once again there are a few semi-classical orthogonal polynomials for which

the recurrence coefficients satisfy Painlevé V and for which the required solution is the special function solution. From the integral representation [123,
Eq. 13.4.1]

$$M(a,b,z) = \frac{1}{\Gamma(a)\Gamma(b-a)} \int_0^1 e^{zt} t^{a-1}(1-t)^{b-a-1}\, dt$$

we see that the Toda modification of the Jacobi weight on $[0,1]$

$$w(x,t) = t^{a-1}(1-t)^{b-a-1} e^{xt},$$

has moments

$$m_0 = \int_0^1 w(x,t)\, dt = \Gamma(a)\Gamma(b-a)M(a,b,t), \quad m_k = \Gamma(a)\Gamma(b-a)\frac{d^k}{dt^i}M(a,b,t),$$

so the Hankel determinant involving the moments of this weight is a Wronskian for the Kummer function $M(a,b,t)$, and the special function solution
corresponds to φ in (6.34) or (6.37) containing only the M-function, i.e., with
$c_2 = 0$. This weight is a minor variation of the modified Jacobi weight that was
considered in Section 5.2, where we indeed saw that Painlevé V was found by
Basor, Chen and Ehrhardt [6].

Another example is related to the integral representation [123, Eq. 13.4.4]

$$U(a,b,z) = \frac{1}{\Gamma(a)} \int_0^\infty e^{-zt} t^{a-1}(1+t)^{b-a-1}\, dt.$$

If we take the weight function

$$w(x) = x^{a-1}(1+x)^{b-a-1} e^{-xt}, \qquad x \in [0,\infty),$$

then the moments are given by

$$m_0 = \Gamma(a)U(a,b,t), \quad m_k = (-1)^k \Gamma(a)\frac{d^k}{dt^k}U(a,b,t),$$

and the Hankel determinant for these moments is a Wronskian of the function
$U(a,b,t)$. The recurrence coefficients for the orthogonal polynomials for this
weight are in terms of such Hankel determinants and satisfy the Painlevé V
equation, where the special function solution in terms of the Kummer function
U are used, i.e., one uses φ from (6.34) or (6.37) with $c_1 = 0$. Chen and Dai
[24] considered a related weight

$$w(x,t) = x^\alpha (1-x)^\beta e^{-t/x}, \qquad x \in [-1,1],$$

which they called a Pollaczek–Jacobi type weight. The moments are given by

$$m_k = \int_0^1 x^{\alpha+k}(1-x)^\beta e^{-t/x}\, dx,$$

which after the change of variable $x = 1/y$ gives

$$m_k = e^{-t} \int_0^\infty y^b (y+1)^{-a-b-k-2} e^{-ty} \, dy = \Gamma(b+1) e^{-t} U(b+1, -a-k, t),$$

and hence the Hankel determinant is in terms of the Kummer function $e^{-t}U$, which is the φ from (6.37). The recurrence coefficients of the corresponding orthogonal polynomials indeed satisfy Painlevé V, as was shown in [24].

A third example are the discrete orthogonal polynomials with the generalized Meixner weights

$$w_k = \frac{(\gamma)_k a^k}{(\beta)_k k!}, \qquad k \in \mathbb{N},$$

which we considered in Section 5.3. The moments are

$$m_0 = \sum_{k=0}^\infty \frac{(\gamma)_k a^k}{(\beta)_k k!} = M(\gamma, \beta, a), \qquad m_k = \left(a \frac{d}{da} \right)^k M(\gamma, \beta, a),$$

and hence the Hankel determinant of these moments is a Wronskian type determinant for the Kummer function $M(\gamma, \beta, a)$ (in the variable a). The recurrence coefficients indeed satisfy the Painlevé V equation (Clarkson [41, §5.2]), as was mentioned in Section 5.3

6.2.5 Painlevé VI

For P$_{VI}$ in (1.23) there are special function solutions in terms of the Gauss hypergeometric function $F(a, b; c; z) = {}_2F_1(a, b; c; z)$, see [123, §15]. These special solutions appear if and only if

$$a + b + c + d = 2n + 1, \qquad n \in \mathbb{Z},$$

where $a = \pm\sqrt{2\alpha}$, $b = \pm\sqrt{-2\beta}$, $c = \pm\sqrt{2\gamma}$, $d = \pm\sqrt{1 - 2\delta}$, where the $\pm$ sign can be taken independently [39, Thm. 7.7], [82, §44]. The Riccati equation is given by

$$x(x-1)y' = ay^2 + [(b+c)x - a - c]y - bx, \tag{6.38}$$

for which the solutions (after a lengthy but straightforward computation) are also solutions of P$_{VI}$ when $a + b + c + d = 1$. To find the solution of (6.38) we set

$$y(x) = \frac{z-1}{a} \frac{\varphi'(z)}{\varphi(z)}, \qquad z = \frac{1}{1-x},$$

and then the function φ has to satisfy the second order linear differential equation

$$z(1-z)\varphi'' + [b + c - (1 - a + b)z]\varphi' + ab\varphi = 0,$$

which one recognizes as the hypergeometric equation. Comparing the parameters with the standard hypergeometric equation [123, Eq. 15.10.1], we find that the general solution is

$$\varphi(z) = c_1 F(b, -a; b + c; z) + c_2 z^{1-b-c} F(1 - c, 1 - a - b - c; 2 - b - c; z). \quad (6.39)$$

Using the corresponding $y_0(x)$ as the seed solution, the Bäcklund transformations [123, Eq. 32.7.46] will give the special function solutions for $a+b+c+d = 2n + 1$. These solutions have a determinantal expression in terms of Wronskian determinants for the function φ in (6.39), see [133], [73], [117].

There are semi-classical orthogonal polynomials that have recurrence coefficients which are in terms of these special function solutions. The integral representation [123, Eq. 15.6.1]

$$F(a, b; c; t) = \frac{1}{\Gamma(b)\Gamma(c - b)} \int_0^1 \frac{x^{b-1}(1 - x)^{c-b-1}}{(1 - tx)^a} \, dx, \qquad \Re c > \Re b > 0,$$

shows that the modified Jacobi weight function

$$w(x, t) = \frac{x^{b-1}(1 - x)^{c-b-1}}{(1 - xt)^a}, \qquad x \in [0, 1],$$

has moments

$$m_0 = \Gamma(b)\Gamma(c - b)F(a, b; c; t), \qquad m_k = \Gamma(b + k)\Gamma(c - b)F(a, b + k; c + k; t),$$

and that

$$\frac{d^k m_0}{dt^t} = \frac{\Gamma(a + k)\Gamma(b + k)\Gamma(c - b)}{\Gamma(a)} F(a + k, b + k; c + k; t).$$

Dai and Zhang [49] and Lyu and Chen [110] investigated the weight function

$$w(x, t) = x^\alpha (1 - x)^\beta |x - t|^\gamma, \qquad x \in [0, 1],$$

and they indeed found that Painlevé VI is the relevant equation for this family of orthogonal polynomials. Chen and Zhang [28] considered the weight function

$$w(x, t) = x^\alpha (1 - x)^\beta (A + B\Theta(x - t)), \qquad x \in [0, 1],$$

where Θ is the Heaviside function. The moments can be expressed in terms of hypergeometric functions, and the Painlevé VI equation, with its special solution in terms of hypergeometric functions, is appearing for these orthogonal polynomials.

A discrete semi-classical weight with hypergeometric moments has weights

$$w_k = \frac{(\alpha)_k(\beta)_k a^k}{(\gamma)_k k!}, \qquad k \in \mathbb{N},$$

or its variant on the shifted lattice $\mathbb{N} + 1 - \gamma$. Here we need $0 < a < 1$. The moments on the lattice $\mathbb{N}$ are

$$m_0 = F(\alpha, \beta; \gamma; a), \quad m_k = \left(a \frac{d}{da}\right)^k m_0,$$

so the Hankel matrix containing the moments is a Wronskian type determinant for the differential operator $\delta = a \frac{d}{da}$. The recurrence coefficients therefore should satisfy Painlevé VI. This discrete weight has not been investigated so far, or at least we were not able to find it in the literature.

7

Asymptotic behavior of orthogonal polynomials near critical points

Various special functions are needed to describe the asymptotic behavior of orthogonal polynomials as the degree n tends to infinity. As an example, consider the Laguerre polynomials L_n^α which for $\alpha > -1$ are orthogonal on $[0, \infty)$ with respect to the gamma density:

$$\int_0^\infty L_n^\alpha(x) L_m^\alpha(x) x^\alpha e^{-x}\, dx = 0, \qquad n \neq m.$$

The zeros of the Laguerre polynomials are dense on $[0, \infty)$ and the largest zero is of the order $4n$. The first thing one needs to do when one studies the asymptotic behavior of orthogonal polynomials is to find the asymptotic distribution of the zeros. Let $x_{1,n} < x_{2,n} < \ldots < x_{n,n}$ be the zeros of L_n^α, then in order to prevent the zeros from going to infinity one considers the scaled zeros $x_{j,n}/n$. The asymptotic distribution of the scaled zeros is given by

$$\lim_{n \to \infty} \frac{1}{n} \sum_{j=1}^n f\left(\frac{x_{j,n}}{n}\right) = \frac{1}{2\pi} \int_0^4 f(x) \sqrt{\frac{4-x}{x}}\, dx,$$

for every continuous function f on $[0, 4]$. The limit distribution on the right-hand side is known as the Marchenko–Pastur distribution (or in fact a special case of it). Note that the density $\frac{1}{2\pi}\sqrt{\frac{4-x}{x}}$ vanishes at $x = 4$ as the square root, i.e., $\approx (4-x)^{1/2}$, and at the origin $x = 0$ it tends to infinity as a square root, i.e., $\approx x^{-1/2}$. In the literature one calls the endpoint 4 a *soft edge* and the endpoint 0 a *hard edge*. Zeros cannot go beyond the hard edge 0 because the support $[0, \infty)$ of the gamma distribution forces them to stay to the right of 0. The soft edge however is not enforced by the support of the gamma distribution, but rather by the exponential rate at which the gamma density tends to zero at infinity. The Marchenko–Pastur distribution is the solution of an equilibrium

115

problem in logarithmic potential theory. Let

$$I(\mu) = \int \int \log \frac{1}{|x-y|} \, d\mu(x) \, d\mu(y)$$

be the logarithmic energy of a probability measure μ. The minimum of

$$I(\mu) + \int_0^\infty x \, d\mu(x) \tag{7.1}$$

for all probability measures μ supported on $[0, \infty)$ is given by the Marchenko–Pastur distribution. The support $[0, 4]$ is a subset of $[0, \infty)$ and the endpoint 4 (the soft edge) arises because of the second term in (7.1).

For the asymptotic behavior of L_n^α (as $n \to \infty$) one needs to distinguish four regions in the complex plane (see, e.g., [146, Thm. 8.22.8]):

1. Inside the interval $(0, 4)$ where the zeros are, which is the oscillatory region, one has

$$e^{-x/2}(-1)^n L_n^\alpha(x)$$
$$= \frac{n^{\alpha/2-1/4}}{\sqrt{\pi \sin \phi} x^{\alpha/2+1/4}} \left\{ \sin\left[(n + \frac{\alpha+1}{2})(\sin 2\phi - 2\phi) + \frac{3\pi}{4}\right] + O(\frac{1}{n}) \right\},$$

   where $x = (4n + 2\alpha + 2)\cos^2 \phi$ and the O term holds uniformly when $\epsilon \leq \phi \leq \frac{\pi}{2} - \frac{\epsilon}{\sqrt{n}}$.

2. Outside the interval $[0, 4]$ where the polynomials have exponential growth, one has

$$e^{-x/2}(-1)^n L_n^\alpha(x)$$
$$= \frac{1}{2}(-1)^n \frac{n^{\alpha/2-1/4}}{\sqrt{\pi \sinh \phi} x^{\alpha/2+1/4}} \exp\left[(n + \frac{\alpha+1}{2})(2\phi - \sinh 2\phi)\right]\left(1 + O(\frac{1}{n})\right),$$

   where $x = (4n + 2\alpha + 2)\cosh^2 \phi$ and the O term holds uniformly for $\epsilon \leq \phi \leq \omega$.

3. Near the hard edge $x = 0$ the polynomial changes from oscillatory behavior to monotonic behavior and one needs the Bessel function J_α to describe the asymptotic behavior:

$$e^{-x/2} x^{\alpha/2} L_n^\alpha(x) = N^{-\alpha/2} \frac{\Gamma(n + \alpha + 1)}{n!} J_\alpha(2(Nx)^{1/2}) + O(n^{\alpha/2-3/4}),$$

   where $N = n + \frac{\alpha+1}{2}$. Typically this is useful when $x = y/N$, hence for x close to 0. Note that the order of the Bessel function is in terms of the behavior of the gamma density near 0.

4. Near the soft edge $x = 4$ the polynomial also changes from oscillatory behavior to monotonic behavior, but now one needs the Airy function Ai to describe the asymptotic behavior:

$$e^{-x/2}L_n^\alpha(x) = (-1)^n \frac{1}{2^{\alpha+1/3}n^{1/3}}\big(\mathrm{Ai}(t) + O(n^{-2/3})\big),$$

where $x = 4n + 2\alpha + 2 + 2(2n)^{1/3}t$ and the O term holds uniformly when t is bounded.

These formulas are known as Plancherel–Rotach formulas, except for the formula near the origin which is known as Hilb's formula. The special functions that we need for describing the asymptotic behavior are the trigonometric functions sin and cos, the exponential function and the hyperbolic functions cosh and sinh, the Bessel functions J_α and the Airy function Ai.

In random matrix theory one is often interested in the asymptotic behavior of the Christoffel–Darboux kernel

$$K_n(x, y) = \sum_{k=0}^{n-1} p_k(x)p_k(y)$$

for which we know the Christoffel–Darboux formula [146, §3.2]

$$K_n(x, y) = a_n \frac{p_n(x)p_{n-1}(y) - p_{n-1}(x)p_n(y)}{x - y},$$

where a_n is the recurrence coefficient (1.4). The reason is that for the Gaussian Unitary Ensemble (GUE) with probability density

$$\frac{1}{Z_n}e^{-\mathrm{Tr}\,V(M)}\,dM$$

on Hermitian $n \times n$ matrices M, the density of the eigenvalues is given by

$$P_n(x_1, x_2, \ldots, x_n) = \frac{1}{n!}\det[\widetilde{K}_n(x_i, x_j)]_{i,j=1}^n,$$

with

$$\widetilde{K}_n(x, y) = K_n(x, y)\sqrt{w(x)w(y)}, \tag{7.2}$$

where $w(x) = e^{-V(x)}$, and the k-point correlation functions

$$R_k(x_1, \ldots, x_k) = \int \cdots \int P_n(x_1, x_2, \ldots, x_n)\,dx_{k+1}\cdots dx_n$$

are given by a similar expression

$$R_k(x_1, \ldots, x_k) = \det[\widetilde{K}_n(x_i, x_j)]_{i,j=1}^k.$$

In particular one finds that the average density of the eigenvalues is

$$\frac{1}{n}R_1(x) = \frac{1}{n}\widetilde{K}_n(x, x) = \frac{1}{n}K_n(x, x)w(x).$$

The famous universality result in random matrix theory is, in terms of these Christoffel–Darboux kernels, a result on the behavior of the Christoffel–Darboux kernel, i.e., an asymptotic result of the form

$$\lim_{n\to\infty} \frac{1}{n^c} \widetilde{K}_n(x^* + s/n^\gamma, x^* + t/n^\gamma) = K(s, t),$$

for some c, γ, where the limit is independent of the point x^* and of the particular potential V in the weight function. Typical limit kernels are

$$K_{\sin}(s, t) = \frac{\sin(s - t)}{s - t}, \tag{7.3}$$

which appears when x^* is in the bulk of the spectrum (the set where the eigenvalues accumulate),

$$K_{Ai}(s, t) = \frac{Ai(s)Ai'(t) - Ai'(s)Ai(t)}{s - t}, \tag{7.4}$$

containing the Airy function, and which appears when x^* is on the soft edge of the spectrum, and

$$K_{Bessel}^\nu(s, t) = \frac{\sqrt{st}}{2} \frac{J_\nu(s)J_{\nu-1}(t) - J_{\nu-1}(s)J_\nu(t)}{s - t}, \tag{7.5}$$

with the Bessel function J_ν, which appears when x^* is on the hard edge of the spectrum.

Nonlinear special functions, and in particular Painlevé transcendents, are needed when the support of the asymptotic zero distribution has special points and the behavior of the density of the asymptotic zero distribution has a behavior different from square root behavior at a hard edge or a soft edge. Such special points are known as critical points and the asymptotic behavior at such points, or when one moves towards such points at a particular rate, has only been studied since the 1990s. The main reason is that tools became available around that time that allow to do the asymptotic analysis rigorously, namely the Riemann–Hilbert problem for orthogonal polynomials (introduced by Fokas, Its and Kitaev [69] in 1992) and the steepest descent method for oscillatory Riemann–Hilbert problems (introduced by Deift and Zhou [54] in 1993). We will discuss some of these critical points and their related Painlevé transcendents in the following sections. See also the short survey of Dan Dai [47], which was very useful in the preparation of this chapter.

7.1 Painlevé I

In Chapter 2 we saw that the orthogonal polynomials for the Freud weight $e^{-x^4 + tx^2}$ have recurrence coefficients that satisfy d-P_I and we gave the asymptotic behavior of these recurrence coefficients. Let us consider the slightly different weight function $w(x) = e^{-N(x^2/2 + tx^4/4)}$, where N is a positive integer. For $t > 0$ the (monic) orthogonal polynomials always exist and they satisfy a three term recurrence relation of the form $xP_n(x) = P_{n+1}(x) + a_n^2 P_{n-1}(x)$, where the recurrence coefficients a_n^2 depend on N and t, so it is really $a_{n,N}^2(t)$ but we drop the N and t when we don't need them. The discrete Painlevé equation for this weight is

$$a_n^2 + ta_n^2(a_{n-1}^2 + a_n^2 + a_{n+1}^2) = \frac{n}{N}.$$

The zeros of the orthogonal polynomials $P_n(x)$ for $N = n$ have an asymptotic zero distribution given by

$$\lim_{n \to \infty} \frac{1}{n} \sum_{j=1}^{n} f(x_{j,n}) = \frac{t}{2\pi} \int_{-c_t}^{c_t} f(x)(x^2 - d_t^2)\sqrt{c_t^2 - x^2}\, dx,$$

for every continuous function f on $[-c_t, c_t]$, with

$$c_t^2 = \frac{2}{3t}(\sqrt{1 + 12t} - 1), \quad d_t^2 = -\frac{1}{3t}(\sqrt{1 + 12t} + 2). \tag{7.6}$$

Hence for $t > 0$ there are two soft edges at $\pm c_t$. The asymptotic behavior of the recurrence coefficients is given by

$$a_{n,n}^2(t) = \frac{c_t^2}{4} + O(\frac{1}{n}). \tag{7.7}$$

If $t = 0$ we get the Hermite polynomials and $td_t^1 \to -1$, $c_t^2 \to 4$. The equilibrium density (the density of the zeros) is the semi-circle density $\frac{1}{2\pi}\sqrt{4 - x^2}$ on $[-2, 2]$. Surprisingly, the discrete Painlevé equation has for $t < 0$ still a solution that satisfies the asymptotic behavior in (7.7), but the weight function is not integrable anymore on the real line. The expressions for c_t^2 and d_t^2 in (7.6) give positive real values for $-\frac{1}{12} < t < 0$, but $t = -1/12$ is a critical value since for $t < -1/12$ the formulas give complex values. Observe that for $t = -1/12$ we find that $c_t^2 = 8 = d_t^2$, so the density of the asymptotic zero distribution becomes

$$\frac{1}{24\pi}(8 - x^2)^{3/2}, \qquad x \in [-\sqrt{8}, \sqrt{8}]. \tag{7.8}$$

This density has a different behavior near its endpoints: instead of a square root vanishing, we now get a 3/2 root vanishing at the endpoints. It turns out that for $t = -1/2$ one still gets $a_{n,n}^2 \to c_t^2/4 = 2$, but the error term is no longer $O(1/n)$, as in (7.7).

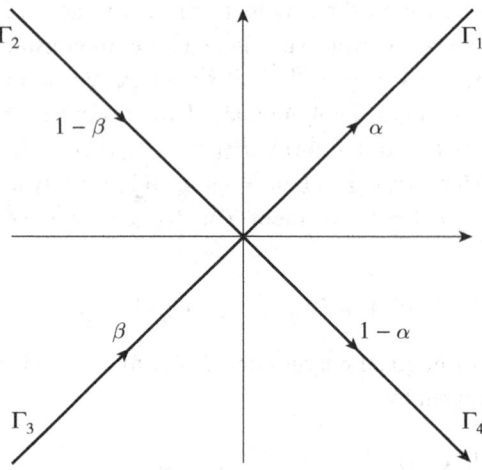

Figure 7.1 The contours for the bilinear form (7.9)

First of all, one needs to make sense of the orthogonality, since obviously integrating over $(-\infty, \infty)$ will not be possible. Fokas, Its and Kitaev [69, 70] suggested to integrate over the lines $\Gamma = \{re^{\pm i\pi/4}, -\infty < r < \infty\}$ in the complex plane, because there the weight function tends to 0 as $r \to \infty$. Duits and Kuijlaars [57] extended this idea a bit and gave weights to the different parts of these curves. Let $\Gamma = \Gamma_1 \cup \Gamma_2 \cup \Gamma_3 \cup \Gamma_4$, where $\Gamma_k = \{re^{(2k-1)i\pi/4}, 0 \le r < \infty\}$, and consider the bilinear form

$$\langle p, q \rangle = \sum_{k=1}^{4} \alpha_k \int_{\Gamma_k} p(z)q(z)e^{-N(z^2/2 + tz^4/4)} \, dz, \tag{7.9}$$

where $\alpha_1 = \alpha, \alpha_3 = \beta$ and $\alpha_2 = 1-\beta, \alpha_4 = 1-\alpha$. One can then ask for the monic orthogonal polynomials, for which $\langle P_n, z^k \rangle = 0$, for $0 \le k \le n-1$. The bilinear form is not Hermitian, so the existence of these orthogonal polynomials is not guaranteed. However, when they exist, they will still satisfy a three term recurrence relation, but since we lost the symmetry, it will be of the form

$$P_{n+1}(z) = (z - b_n)P_n(z) - a_n^2 P_{n-1}(z).$$

Fokas, Its and Kitaev [69, 70] and Duits and Kuijlaars [57] proved the following result (I used the formulation in [57, Thm. 1.2]).

Theorem 7.1 *Let α, β be complex numbers and let t tend to $-1/12$ in such a way that*

$$n^{4/5}(t + 1/12) = -c_1 x$$

remains fixed. Then for large enough n the recurrence coefficients $a_{n,n}(t)$ and $b_{n,n}(t)$ for the orthogonal polynomials P_n exist and

$$a_{n,n}(t) = 2 - c_2(y_\alpha(x) + y_\beta(x))n^{-2/5} + O(n^{-3/5}),$$

and

$$b_{n,n}(t) = c_3(y_\beta(x) - y_\alpha(x))n^{-2/5} + O(n^{-3/5}),$$

as $n \to \infty$, with specific values of the constants c_1, c_2, c_3[1], where y_α and y_β are solutions of the Painlevé I equation $y'' = 6y^2 + x$. The asymptotics hold uniformly for x in compact subsets of $\mathbb{R}$ not containing any of the poles of y_α or y_β.

All the y_α are solutions of the P_I equation for which

$$y(x) = \sqrt{-x/6}(1 + o(1)), \qquad x \to -\infty,$$

and have a common asymptotic series

$$y_\alpha(x) \sim \sqrt{-x/6}\left(1 + \sum_{k=1}^{\infty} a_k(-x)^{-5k/2}\right), \qquad x \to -\infty.$$

For $\alpha = 1$ the asymptotic series is valid for $|x| \to \infty$ in the region $\arg x \in [\frac{3\pi}{5}, \pi]$, and the other solutions y_α differ from y_1 only by exponentially small terms:

$$y_\alpha(x) = y_1(x) - \frac{i(\alpha - 1)}{\sqrt{\pi} 2^{11/8} 3^{1/8} (-x)^{1/8}} e^{-\frac{1}{5} 2^{11/4} 3^{1/4} (-x)^{5/4}} (1 + O(x^{-3/8})), \qquad x \to -\infty.$$

The proof is based on the Riemann–Hilbert problem for orthogonal polynomials, which was described in Section 4.2 for weights on the real line. We give a sketch of the proof but refer to [57] for all the details. The Riemann–Hilbert problem still holds in the present situation, except that one now has jump conditions on the contours Γ_k:

$$Y_+(z) = Y_-(z)\begin{pmatrix} 1 & \alpha_k w(z) \\ 0 & 1 \end{pmatrix}, \qquad z \in \Gamma_k, \ k = 1, 2, 3, 4,$$

where the Y_+ are now the limiting values taken from the left side of the curve, and Y_- these taken from the right side of the curve, with left and write indicated by the arrow on the curve. In the solution Y one uses integrals over the contour

---

[1] One has $c_1 = 2^{-9/5} 3^{-6/5}$, $c_2 = 2^{3/5} 3^{2/5}$, $c_3 = 2^{1/10} 3^{2/5}$, if you really want to know.

Γ with the appropriate weights in the second column. The Riemann–Hilbert problem has a solution if and only if the monic orthogonal polynomials P_n and P_{n-1} and the constant γ_{n-1}^2 exist. In order to obtain the asymptotic behavior, one transforms the Riemann–Hilbert problem to one that can easily be analyzed for $n \to \infty$. The first transformation is to modify the contour Γ in such a way that the polynomials P_n and P_{n-1} remain unchanged. This can be done because the weight w is analytic, and one only needs to make sure that it tends to zero when the variable tends to infinity on the contour. The new contour is of the form $\Gamma_0 \cup \Gamma_1 \cup \Gamma_2 \cup \Gamma_3 \cup \Gamma_4$, where $\Gamma_0 = [-c_t, c_t]$, and the other Γ_k connect the endpoints $\pm c_t$ with $e^{+\infty(2k-1)i\pi/4}$, see Figure 7.2.

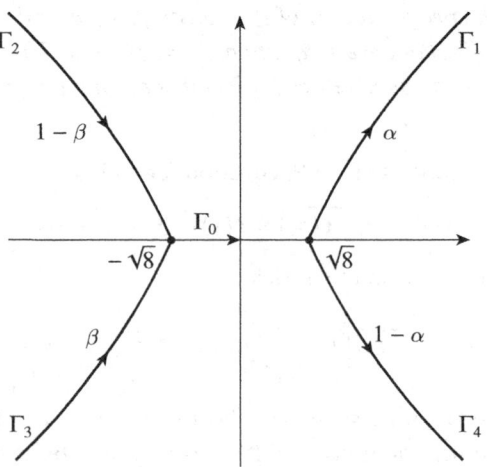

Figure 7.2 The deformed contour

The new Γ_k with $k = 1, 2, 3, 4$ keep the same weights $\alpha_1, \alpha_2, \alpha_3, \alpha_4$, and for Γ_0 we take the weight $\alpha_0 = 1$. The bilinear form now is

$$\langle p, q \rangle = \sum_{k=0}^{4} \alpha_k \int_{\Gamma_k} p(z)q(z)e^{-N(z^2/2+tz^4/4)} \, dz. \qquad (7.10)$$

By Cauchy's theorem the bilinear form doesn't change when p and q are polynomials. Let

$$\phi(z) = \frac{1}{24} \int_{\sqrt{8}}^{z} (s^2 - 8)^{3/2} \, ds, \qquad (7.11)$$

then $\phi(\sqrt{8}) = 0$, $\phi(-\sqrt{8}) = \pi$, hence $\Im\phi$ vanishes at $\pm\sqrt{8}$. Take for $\Gamma_1, \Gamma_2, \Gamma_3, \Gamma_4$ curves for which $\Im\phi = 0$, and in particular the steepest descent curves for which $\Re\phi$ tends to $-\infty$.

The next transformation is to normalize the Riemann–Hilbert problem: we want to have a Riemann–Hilbert problem that for $z \to \infty$ converges to the identity matrix $\mathbb{I}$. For this one uses the g-function

$$g_t(z) = \int_{-\sqrt{8}}^{\sqrt{8}} \log(z - x) \, dv_t(x),$$

where v_t is the *signed measure* with support in $[-\sqrt{8}, \sqrt{8}]$ that minimizes

$$\int \int \log \frac{1}{|x - y|} \, dv(x) \, dv(y) + \int \left(\frac{x^2}{2} + t\frac{x^4}{4} \right) dv(x)$$

among all signed measures v with $\mathrm{supp}(v) \subset [-\sqrt{8}, \sqrt{8}]$ and $v([-\sqrt{8}, \sqrt{8}]) = 1$. This is not the measure in (7.8), unless $t = -1/2$. It turns out that $g_t(z) = \log z + O(1/z)$ as $z \to \infty$ and

$$g_{t+}(z) + g_{t-}(z) = \frac{z^2}{2} + t\frac{z^4}{4} + \ell_t, \qquad z \in (-\sqrt{8}, \sqrt{8}),$$

where ℓ_t is a constant. We then transform the Riemann–Hilbert matrix Y to a new matrix T by

$$T(z) = \begin{pmatrix} e^{-n\ell_t/2} & 0 \\ 0 & e^{n\ell_t/2} \end{pmatrix} Y \begin{pmatrix} e^{-ng_t(z)} & 0 \\ 0 & e^{ng_t(z)} \end{pmatrix} \begin{pmatrix} e^{n\ell_t/2} & 0 \\ 0 & e^{-n\ell_t/2} \end{pmatrix}.$$

This new matrix T now has jumps on the curves given by

$$T_+(z) = T_-(z) \begin{pmatrix} 1 & \alpha_k e^{2n\phi_t(z)} \\ 0 & 1 \end{pmatrix}, \qquad z \in \Gamma_k, \ k = 1, 2, 3, 4,$$

and

$$T_+(z) = T_-(z) \begin{pmatrix} e^{-2n\phi_{t+}(z)} & 0 \\ 0 & e^{-2n\phi_{t-}(z)} \end{pmatrix}, \qquad z \in (-\sqrt{8}, \sqrt{8}),$$

and it is normalized near infinity: $T(z) = \mathbb{I} + O(1/z)$. The function ϕ_t is given by

$$\phi_t(z) = -\frac{1}{2}\left(\frac{z^2}{2} + t\frac{z^4}{4} \right) + g_t(z) - \frac{\ell_t}{2},$$

and it is closely related to the function ϕ in (7.11) because $\phi_t = \phi + (t + 1/12)\phi^0$, for a certain explicit function ϕ^0 [57, Eq. (4.12)].

The next step is to deform the contour Γ_0. It turns out that $\Re \phi_t(z)$ is purely imaginary for $z \in (-\sqrt{8}, \sqrt{8}) = \Gamma_0$, so that the jump of T on Γ_0 is oscillatory. The Deift–Zhou steepest descent method then consists of opening a lens around Γ_0 (see Figure 7.3) such that the jump over Γ_0 is distributed over three jumps over the two lips of the lens and Γ_0, but in such a way that the jumps on the lips converge to the identity matrix $\mathbb{I}$ as $n \to \infty$ and the jump over Γ_0 is independent of n. This can be done because the jump matrix for Γ_0 can be factored as a product of three matrices.

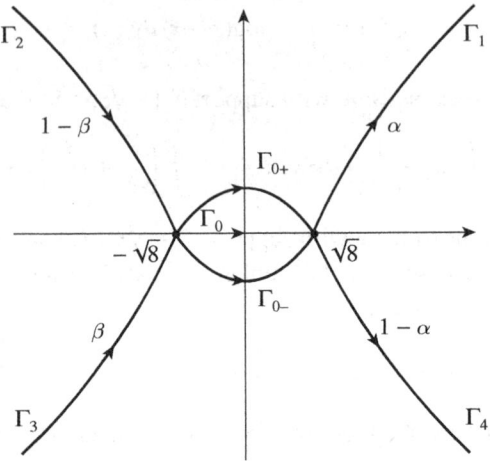

Figure 7.3 Opening the lens around Γ_0

The new Riemann–Hilbert matrix will be

$$S(z) = T(z) \begin{pmatrix} 1 & 0 \\ -e^{-2n\phi_t(z)} & 1 \end{pmatrix}, \qquad \text{between } \Gamma_0 \text{ and } \Gamma_{0,+},$$

$$S(z) = T(z) \begin{pmatrix} 1 & 0 \\ e^{-2n\phi_t(z)} & 1 \end{pmatrix}, \qquad \text{between } \Gamma_0 \text{ and } \Gamma_{0,-},$$

and $S(z) = T(z)$ elsewhere. This matrix has the same jumps as T on $\Gamma_1, \Gamma_2, \Gamma_3$ and Γ_4, and new jumps

$$S_+(z) = S_-(z) \begin{pmatrix} 1 & 0 \\ e^{-2n\Phi_t(z)} & 1 \end{pmatrix}, \qquad z \in \Gamma_{0+} \cup \Gamma_{0-},$$

and

$$S_+(z) = S_-(z) \begin{pmatrix} 0 & 1 \\ -1 & 0 \end{pmatrix}, \qquad z \in \Gamma_0 = (-\sqrt{8}, \sqrt{8}).$$

What is pleasing now is that S has jumps on $\Gamma_1, \Gamma_2, \Gamma_3, \Gamma_4, \Gamma_{0+}, \Gamma_{0-}$ that converge to the identity matrix as $n \to \infty$, which follows because $\Re\phi_t < 0$ on those curves, away from $\pm \sqrt{8}$. It implies that the main contribution for the matrix S is the matrix M which satisfies the Riemann–Hilbert problem

1. M is analytic on $\mathbb{C} \setminus [-\sqrt{8}, \sqrt{8}]$,

2. M has the jump condition

$$M_+(x) = M_-(x) \begin{pmatrix} 0 & 1 \\ -1 & 0 \end{pmatrix}, \qquad x \in (-\sqrt{8}, \sqrt{8}),$$

3. $M(z) = \mathbb{I} + O(1/z)$ as $z \to \infty$.

This matrix M is called the *global parametrix* and this Riemann–Hilbert problem has a solution which was already used several times in the literature (see, e.g., Deift's monograph [51]):

$$M(z) = \frac{1}{2} \begin{pmatrix} 1 & 1 \\ i & -i \end{pmatrix} \begin{pmatrix} m(z) & 0 \\ 0 & 1/m(z) \end{pmatrix} \begin{pmatrix} 1 & -i \\ 1 & i \end{pmatrix}, \qquad m(z) = \left(\frac{z - \sqrt{8}}{z + \sqrt{8}} \right)^{1/4}.$$

The matrix SM^{-1} does not have a jump any longer on $(-\sqrt{8}, \sqrt{8})$, the jumps on the remaining curves tend to the identity matrix, and it is close to the identity matrix as $z \to \infty$. One would like to deduce from this that SM^{-1} therefore converges to the identity matrix everywhere in the complex plane. Unfortunately, the necessary requirement for such a conclusion is that the jumps of SM^{-1} have to converge to the identity matrix *uniformly* on the remaining curves, and this is not the case in the neighborhood of the points $\pm \sqrt{8}$. Hence one needs to make a local analysis around these two points. What is needed is to find explicit Riemann–Hilbert problems which have the jumps of S in a neighborhood $\Omega_\pm$ of $\pm \sqrt{8}$ and which match the global parametrix M on the boundary of those neighborhoods with an error $O(n^{-c})$ for some $c > 0$. These explicit Riemann–Hilbert problems are the *local parametrices* around these points. Usually these local parametrices are in terms of the Airy function (when the density of the asymptotic zero distribution vanishes like $|x-a|^{1/2}$ around the point a) or Bessel functions (when the density is unbounded and tends to infinity like $|x - a|^{-1/2}$), but in the case under investigation, the density (7.8) tends to zero at the endpoints with a power $3/2$, and this requires a different parametrix.

Let us see what happens near $\sqrt{8}$, the analysis around $-\sqrt{8}$ is similar. The Riemann–Hilbert problem for S near $\sqrt{8}$ is on 5 curves that meet at $\sqrt{8}$ (see Figure 7.4).

We can get rid of the $e^{\pm 2n\phi_t(z)}$ in these jumps by multiplying with an appropriate diagonal matrix, so what is left is a Riemann–Hilbert problem with constant jumps over 5 lines meeting at $\sqrt{8}$. This is where Painlevé I comes in, because there is a nice Riemann–Hilbert problem for Painlevé I, formulated by Kapaev [99] in 2004, see also [68, §5.2]. The Riemann–Hilbert problem is on 5 rays emanating from the origin at angles $2k\pi/5$, $k \in \{-2, -1, 0, 1, 2\}$, together with the negative axis. On each of the rays there are constant jump matrices,

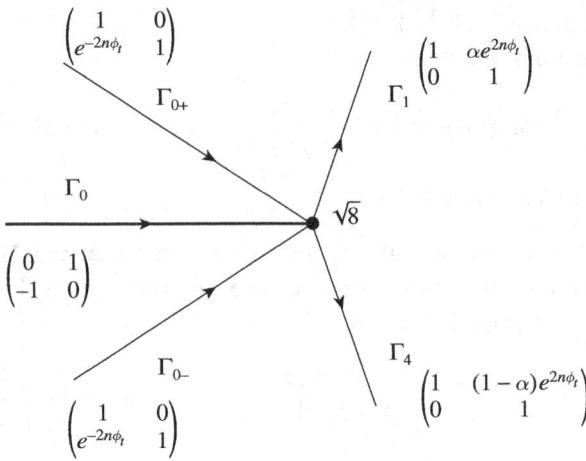

Figure 7.4 Local behavior of S near $\sqrt{8}$

as indicated in Figure 7.5, where $s_{-2}, s_{-1}, s_0, s_1, s_2$ are *Stokes multipliers* satisfying

$$1 + s_k s_{k+1} = -i s_{k+3}, \qquad s_{k+5} = s_k, \qquad k \in \mathbb{Z}.$$

The Riemann–Hilbert problem for this system of curves Σ and jump matrices A_k is to find a 2×2 matrix valued function Ψ^0, depending on a parameter x, for which

- $\Psi^0(\cdot; x)$ is analytic in $\mathbb{C} \setminus \Sigma$,
- $\Psi^0_+(z; x) = \Psi^0_-(z; x) A_k$ for z on each of the rays Σ_k in Σ,
- Asymptotic behavior: for $z \to \infty$

$$\Psi^0(z; x) = \frac{1}{\sqrt{2}} \begin{pmatrix} z^{1/4} & 0 \\ 0 & z^{-1/4} \end{pmatrix} \begin{pmatrix} 1 & 1 \\ 1 & -1 \end{pmatrix} \left(\mathbb{I} + O(1/z^{1/2}) \right) \begin{pmatrix} e^{\theta(z,x)} & 0 \\ 0 & e^{-\theta(z,x)} \end{pmatrix},$$

where $\theta(z, x) = \frac{4}{5} z^{5/2} + x z^{1/2}$.

The matrix Ψ^0 has an expansion in powers of $z^{-1/2}$ of the form

$$\Psi^0(z; x) = \frac{1}{\sqrt{2}} \begin{pmatrix} z^{1/4} & 0 \\ 0 & z^{-1/4} \end{pmatrix} \begin{pmatrix} 1 & 1 \\ 1 & -1 \end{pmatrix} \left(\mathbb{I} + \begin{pmatrix} -H & 0 \\ 0 & H \end{pmatrix} z^{-1/2} \right.$$

$$\left. + \frac{1}{2} \begin{pmatrix} H^2 & y \\ y & H^2 \end{pmatrix} z^{-1} + O(z^{-3/2}) \right) \begin{pmatrix} e^{\theta(z,x)} & 0 \\ 0 & e^{-\theta(z,x)} \end{pmatrix},$$

where H and y depend on x. Then $y = y(x)$ is a solution of the Painlevé I

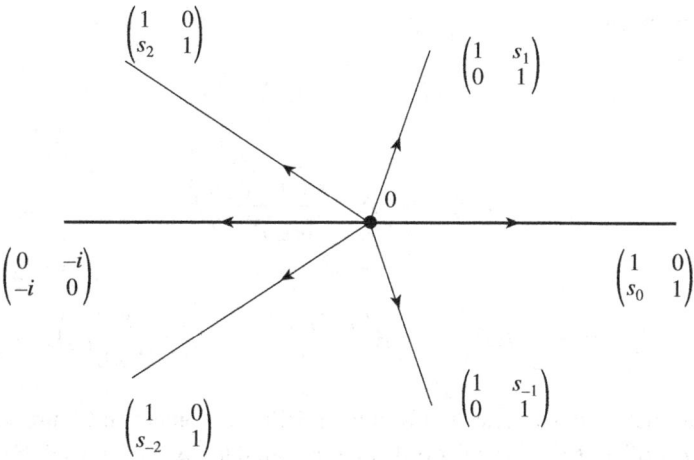

Figure 7.5 The Riemann–Hilbert problem for Ψ^0

equation (1.18) and

$$H(x) = \frac{1}{2}(y'(x))^2 - 2y^3(x) - xy(x).$$

The Riemann–Hilbert problem for Ψ^0 has a unique solution for each set of Stokes multipliers if and only if x is not a pole of y. Our local parametrix corresponds to the choice

$$s_0 = 0, \quad s_1 = i\alpha, \quad s_{-1} = i(1 - \alpha), \quad s_{-2} = s_2 = i,$$

and we call the corresponding solution $\Psi^0(z; x, \alpha)$, and the corresponding solution of P_I is denoted by y_α. Note that the orientation of some of the contours is different from what we need and the jumps are not quite the ones we need. This is fixed by considering

$$\Psi(z; x, \alpha) = \Psi^0(z; x, \alpha) \begin{pmatrix} 1 & 0 \\ 0 & -i \end{pmatrix}.$$

The local parametrix is then given by

$$P_\alpha(z) = E(z)\Psi(n^{2/5} f(z); n^{4/5} u_t(z), \alpha) \begin{pmatrix} e^{-n\phi_t} & 0 \\ 0 & e^{n\phi_t} \end{pmatrix}, \tag{7.12}$$

where f is a conformal map from the neighborhood Ω_+ of $\sqrt{8}$ to a neighborhood of 0, u_t is analytic on Ω and $n^{4/5} u_t(\Omega_+)$ does not contain any poles of y_α, and E is an analytic 2×2 matrix valued function on Ω_+. These functions are explicitly given by

$$f(z) = \left(\frac{5}{4}\phi(z)\right)^{2/5},$$

$$u_t(z) = (4/5)^{1/5}\frac{\phi_t(z) - \phi(z)}{(\phi(z))^{1/5}},$$

$$E(z) = \frac{1}{\sqrt{2}}M(z)\begin{pmatrix} 1 & 1 \\ i & -i \end{pmatrix}\begin{pmatrix} [n^{2/5}f(z)]^{-1/4} & 0 \\ 0 & [n^{2/5}f(z)]^{1/4} \end{pmatrix}.$$

This parametrix P_α satisfies the Riemann–Hilbert problem in Figure 7.4 with $P(z) = M(z)(\mathbb{I}+O(n^{-1/5}))$ uniformly on the boundary $\partial\Omega_+$ of the neighborhood Ω_+ of $\sqrt{8}$.

One can find a similar parametrix P_β in a neighborhood Ω_- of $-\sqrt{8}$. To finalize the analysis, one considers the matrix function R given by

$$R(z) = \begin{cases} S(z)M(z)^{-1}, & z \text{ outside the neighborhoods } \Omega_\pm \text{ of } \pm\sqrt{8}, \\ S(z)P_\alpha^{-1}, & z \in \Omega_+, \\ S(z)P_\beta^{-1}, & z \in \Omega_-. \end{cases}$$

This R solves a Riemann–Hilbert problem on a system of contours given in Figure 7.6 which is normalized and for which the jumps on each of the contours tend uniformly to zero as $O(n^{-1/5})$ as $n \to \infty$. The conclusion is that

$$\lim_{n\to\infty} R(z) = \mathbb{I},$$

and then one can undo all the transformations to get the asymptotic behavior of the original matrix Y. From that asymptotic behavior one finally deduces Theorem 7.1, but for this we refer to [57]. Our main intention here was to indicate how Painlevé I enters into the asymptotic analysis.

For orthogonal polynomials on the real line, the density of the zeros can only vanish like $|x - c|^{\frac{4k+1}{2}}$ at $x = c$, with $k \in \mathbb{N}$ (see, e.g., [53], [102]), hence the vanishing as in (7.8) cannot happen for orthogonal polynomials on the real line but only for orthogonal polynomials with a measure supported on curves in the complex plane. Orthogonal polynomials on the real line for which the zero density vanishes like $|x - c|^{\frac{4k+1}{2}}$ can be analyzed asymptotically near the critical point $x = c$ in terms of solutions in the hierarchy of P_I, as was shown by Claeys and Vanlessen [35].

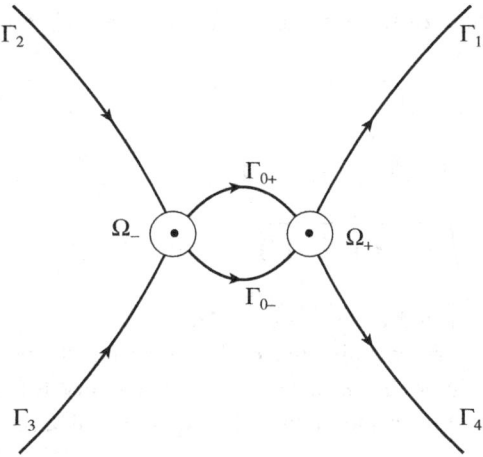

Figure 7.6 The contours for the Riemann–Hilbert problem for R

7.2 Painlevé II

Painlevé II transcendents are typically needed when the density of the zeros of the orthogonal polynomials vanishes quadratically. Bleher and Its [11] investigated the orthogonal polynomials for the Freud weight

$$w_N(x) = e^{-N(x^4/4 + tx^2/2)}, \qquad x \in \mathbb{R}, \tag{7.13}$$

which has a quartic potential $V(x) = x^4/4 + tx^2/2$. We denote the corresponding orthogonal polynomials by $(p_{n,N})_{n \in \mathbb{N}}$, where the first subscript n refers to the degree of the polynomial and the second subscript to the N in the weight function (7.13). Observe that for $t \geq 0$ the potential V has one global minimum on the real axis at the origin, but for $t < 0$ it has two minima at $\pm\sqrt{-t}$. The asymptotic distribution of the zeros of $p_{N,N}$ will be on one interval for $t \geq 0$ and on two intervals (around $\pm\sqrt{-t}$) when $t < 0$ and large, hence there is a phase transition at some critical value $t_c < 0$ where the two intervals touch and become one interval. Other weight functions with a more general potential V also give rise to the same Painlevé II asymptotics: what one needs is that the equilibrium measure is supported on several intervals and two of the intervals are touching, with a quadratic vanishing of the density of the equilibrium measure (see, e.g., [32] and [33]). The same kind of asymptotics also hold for orthogonal polynomials on the unit circle, when the equilibrium measure is supported on one arc and the arc is closing, see e.g., [5]. In this section we explain what happens in the quartic case (7.13), following closely the analysis of Claeys and Kuijlaars [32]. The asymptotic density of the zeros of the corresponding orthogonal polynomials $P_{N,N}$ as $N \to \infty$ is given by

$$
\rho_t(x) = \begin{cases}
\dfrac{x^2 + b_t}{2\pi} \sqrt{a_t^2 - x^2}, & x \in [-a_t, a_t], & \text{if } t > -2, \\[2ex]
\dfrac{x^2}{2\pi} \sqrt{4 - x^2}, & x \in [-2, 2], & \text{if } t = -2, \\[2ex]
\dfrac{|x|}{2\pi} \sqrt{(a_t^2 - x^2)(x^2 - b_t^2)}, & |x| \in [b_t, a_t] & \text{if } t < -2,
\end{cases} \tag{7.14}
$$

where $a_t = \sqrt{2 - t}$ and $b_t = \sqrt{-2 - t}$. Hence there is a phase transition at $t = -2$, which is a critical value where the support of the density of the zeros changes from one interval $[-a_t, a_t]$ to two disjoint intervals $[-a_t, -b_t] \cup [b_t, a_t]$. At $t = -2$ the density vanishes at the origin quadratically, see Figure 7.7.

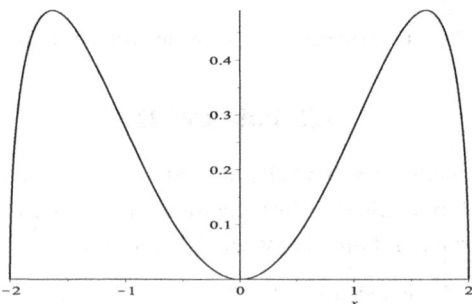

Figure 7.7 The density of the zeros at the critical value $t = -2$

We will investigate the orthogonal polynomials $p_{n,N}$ for n close to N with $n, N \to \infty$ and $n/N \to 1$. The Riemann–Hilbert problem for the monic orthogonal polynomials $P_{n,N}$ starts with the contour $(-\infty, \infty)$ where the jump is in terms of the weight w_N. Then the Riemann–Hilbert problem is normalized at infinity by using g-functions that are defined in terms of the density $\rho_{-2}(x)$ in (7.14). The jump on $(-\infty, \infty)$ now changes to a jump J_0 on $(-\infty, -2)$ and $(2, \infty)$ which converges for $N \to \infty$ uniformly to the identity matrix if we stay away from ± 2, and a jump J_1 on $(-2, 2)$ which is oscillating:

$$
J_0 = \begin{pmatrix} 1 & e^{-2\pi i n \varphi} \\ 0 & 1 \end{pmatrix}, \qquad J_1 = \begin{pmatrix} e^{2\pi i n \varphi} & 1 \\ 0 & e^{-2\pi i n \varphi} \end{pmatrix},
$$

where

$$
\varphi(x) = \int_x^2 \rho(t)\, dt.
$$

The Deift–Zhou steepest descent method for oscillatory Riemann–Hilbert problems can then be used, and this involves opening a lens around $(-2, 2)$ and dis-
tributing the oscillatory jump on $(-2, 2)$ to jumps on the lips of the lens that
are close to the identity matrix and a fixed jump on the interval $(-2, 2)$. Now
due to the quadratic vanishing of the weight ρ_{-2} at the origin, which implies
$\varphi(x) = \frac{1}{2} + O(x^3)$ when $x \to 0$, it is better to open lenses around $(-2, 0)$ and
$(0, 2)$, so that the Riemann–Hilbert problem looks like Figure 7.8.

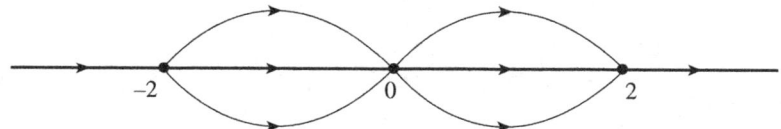

Figure 7.8 Opening the lenses for the Riemann–Hilbert problem

The jumps J_2 on the lips of the lenses are

$$J_2 = \begin{pmatrix} 1 & 0 \\ \Phi^n & 1 \end{pmatrix},$$

with

$$\Phi(z) = \exp\left(2\pi i \int_z^2 \rho_{-2}(t)\, dt \right)$$

and the integral is on a path from z to 2 in the upper or lower complex plane,
depending on whether one is on the upper lip or lower lip of the lens. On those
lips $|\Phi| < 1$, and hence these jumps converge to the identity matrix and the
convergence is uniform if one stays away from ± 2 and 0. In order to control the behavior of the Riemann–Hilbert problem near these points ± 2 and
0, one investigates the Riemann–Hilbert problem locally in a neighborhood of
these points by solving the problem in terms of known functions, but in such a
way that this local solution matches the global solution on the boundary of the
neighborhood. The global solution (global parametrix) is the solution of the
Riemann–Hilbert problem where we ignore all the jumps that converge to the
identity matrix. This Riemann–Hilbert problem only has $[-2, 2]$ as a contour
with a simple jump independent of N. The local solutions are parametrices $P_{\pm 2}$
and P_0 near the points ± 2 and 0, see Figure 7.9.

   The parametrices $P_{\pm 2}$ can be expressed in terms of Airy functions, as is usual
at the soft edge where the density ρ_{-2} vanishes as a square root. The parametrix
P_0 near 0 needs other special functions. If we zoom into the neighborhood of
0, then the Riemann–Hilbert problem looks like Figure 7.10.

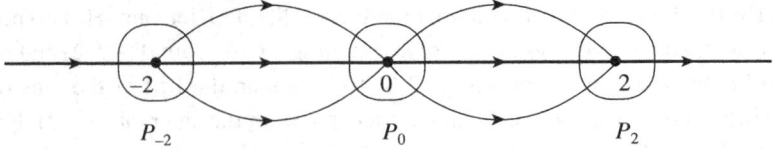

Figure 7.9 Parametrices near ±2 and 0

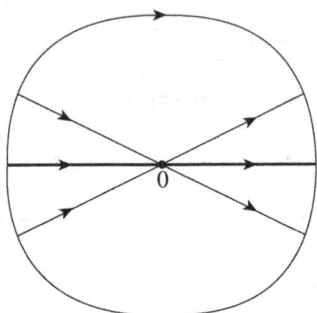

Figure 7.10 The parametrix around 0

We will use the Riemann–Hilbert problem for Painlevé II, which is described in [68, §8.1] and [55]. This is a 2×2 Riemann–Hilbert problem for a matrix $\Psi(\zeta; s)$ with jumps on a family of 6 half lines $\cup_{k=1}^{6} \{re^{i(2k-1)\pi/6}, r \in [0, \infty)\}$, as is shown in Figure 7.11, together with the jump matrices. The asymptotic condition is

$$\Psi(\zeta; s) \begin{pmatrix} e^{i(4\zeta^3/3+s\zeta)} & 0 \\ 0 & e^{-i(4\zeta^3/3+s\zeta)} \end{pmatrix} = \mathbb{I} + O(\zeta^{-1}), \qquad \zeta \to \infty,$$

and near $\zeta = 0$ we require Ψ to be bounded. One then has

$$\frac{d}{d\zeta}\Psi = A\Psi, \qquad \frac{\partial}{\partial s}\Psi = B\Psi,$$

with

$$A = \begin{pmatrix} 4\zeta q(s) & 4\zeta^2 + s + 2q^2 + 2r \\ -4\zeta^2 - s - 2q^2 + 2r & -4\zeta q \end{pmatrix}, \quad B = \begin{pmatrix} q & \zeta \\ -\zeta, -q \end{pmatrix}, \quad (7.15)$$

and the compatibility between these two equations implies that $q = q(s)$ satisfies the Painlevé II equation $q'' = sq + 2q^3$ and $r(s) = q'(s)$.

If we put $s_2 = 0$ then the jump over the vertical axis disappears and the system of contours is similar to the local Riemann–Hilbert problem from Figure 7.10, but without a jump on the real axis, which we will add later. The jumps of

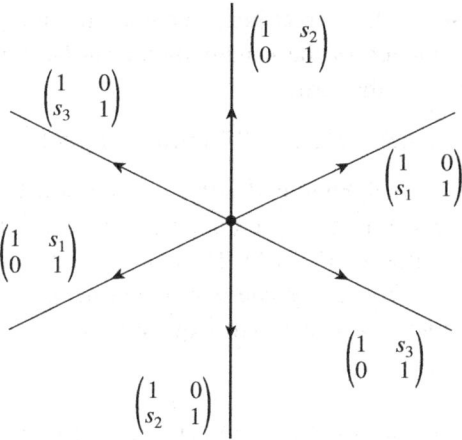

Figure 7.11 Riemann–Hilbert problem for P_{II}

the local parametrix on the other lines match with the jumps for the P_{II} problem if we set $s_1 = 1$ and $s_3 = -1$ and reverse the orientation on the lines $re^{5\pi i/6}$ and $re^{7\pi i/6}$. Recall that the parametrices $P_{\pm 2}$ were such that they match the global parametrix at the boundary of their neighborhoods. This was achieved by using the asymptotic behavior of the Airy functions. We also want to parametrix P_0 to match the global solution at the boundary of its neighborhood, so we need a Painlevé II transcendent that behaves asymptotically like the Airy function. There is indeed such a Painlevé transcendent and this is the Hastings–McLeod solution of P_{II}. This solution corresponds precisely to $s_1 = 1, s_2 = 0, s_3 = -1$ and is characterized by the asymptotic condition

$$q(s) = \text{Ai}(s)(1 + o(1)), \qquad s \to \infty.$$

We just need one more transformation of the Riemann–Hilbert problem for Ψ to get the local parametrix P_0, in particular we need to introduce a jump over the real axis, with jump matrix

$$\begin{pmatrix} 0 & 1 \\ -1 & 0 \end{pmatrix}.$$

Define $M(\zeta; s, \theta)$ by

$$\begin{cases} \begin{pmatrix} e^{i\theta} & 0 \\ 0 & e^{-i\theta} \end{pmatrix} \Psi(\zeta; s) \begin{pmatrix} e^{i(4\zeta^3/3 + s\zeta)} & 0 \\ 0 & e^{-i(4\zeta^3/3 + s\zeta)} \end{pmatrix} \begin{pmatrix} e^{-i\theta} & 0 \\ 0 & e^{i\theta} \end{pmatrix}, & \Im\zeta > 0, \\ \begin{pmatrix} e^{i\theta} & 0 \\ 0 & e^{-i\theta} \end{pmatrix} \Psi(\zeta; s) \begin{pmatrix} e^{i(4\zeta^3/3 + s\zeta)} & 0 \\ 0 & e^{-i(4\zeta^3/3 + s\zeta)} \end{pmatrix} \begin{pmatrix} 0 & -e^{-i\theta} \\ e^{i\theta} & 0 \end{pmatrix}, & \Im\zeta < 0, \end{cases}$$

with $\theta \in \mathbb{R}$, then M has the correct jumps on the lines in Figure 7.10 and the matching on the boundary of the neighborhood can be done explicitly. The local parametrix P_0 is of the form

$$P_0(z) = E(z)M(n^{1/3}f(z); n^{2/3}s(z), n\theta),$$

where E is analytic in a neighborhood of 0, f is a conformal map from a neighborhood Ω around $z = 0$ to a neighborhood of $\zeta = 0$, s is analytic around 0 and θ is a real constant, see [32, §5.6]. The conformal map is essentially constructed using the primitive of the equilibrium density ρ_{-2}, and this primitive behaves like z^3 around $z = 0$, due to the square vanishing of the density ρ_{-2}. The conformal map is

$$f(z) = \left(\frac{3\pi}{4} \int_0^z \rho_{-2}(y)\, dy \right)^{1/3}, \qquad z \in \Omega.$$

This explains the occurrence of many of the third roots and the third power ζ^3. The final result is the following [32, Thm. 2.1], [11]:

Theorem 7.2 *Let $n, N \to \infty$ in such a way that*

$$\lim_{n,N \to \infty} n^{2/3}\left(\frac{n}{N} - 1 \right) = L \in \mathbb{R},$$

then the kernel $\widetilde{K}_{n,N}$ in (7.2) for the weight w_N in (7.13) behaves around the critical point 0 as

$$\lim_{n,N \to \infty} \frac{1}{(cn)^{1/3}} \widetilde{K}_{n,N}\left(\frac{u}{(cn)^{1/3}}, \frac{v}{(cn)^{1/3}} \right) = K_{\mathrm{II}}(u, v; s),$$

uniformly for u and v in compact sets of $\mathbb{R}$, where $c > 0$ and $s \in \mathbb{R}$ are specific constants and

$$K_{\mathrm{II}}(u, v; s) = \frac{\Phi^1(u; s)\Phi^2(v; s) - \Phi^2(u; s)\Phi^1(v; s)}{\pi(u - v)},$$

where Φ^1, Φ^2 are real for real ζ and are a special solution of

$$\frac{d}{d\zeta}\begin{pmatrix} \Phi^1(\zeta; s) \\ \Phi^2(\zeta; s) \end{pmatrix} = A\begin{pmatrix} \Phi^1(\zeta; s) \\ \Phi^2(\zeta; s) \end{pmatrix}, \qquad \frac{\partial}{\partial s}\begin{pmatrix} \Phi^1(\zeta; s) \\ \Phi^2(\zeta; s) \end{pmatrix} = B\begin{pmatrix} \Phi^1(\zeta; s) \\ \Phi^2(\zeta; s) \end{pmatrix},$$

with A and B given in (7.15), with q the Hastings–McLeod solution of Painlevé II and $r = q'(s)$, having the asymptotics

$$\Phi^1(\zeta; s) = \cos\left(\frac{4}{3}\zeta^3 + s\zeta \right) + O(\zeta^{-1}),$$

$$\Phi^2(\zeta; s) = -\sin\left(\frac{4}{3}\zeta^3 + s\zeta \right) + O(\zeta^{-1}),$$

on the real line, as $\zeta \to \pm\infty$.

Its, Kuijlaars and Östensson [93] have investigated random $n \times n$ matrix models for Hermitian matrices with density

$$\frac{1}{Z_{n,N}} |\det M|^{2\alpha} e^{-N \operatorname{Tr} V(M)} \, dM, \qquad \alpha > -\frac{1}{2},$$

for which the relevant orthogonal polynomials have a weight function

$$w(x) = |x|^{2\alpha} e^{-NV(x)}, \qquad x \in \mathbb{R}.$$

This weight function has a singularity at the origin when $\alpha < 0$ or a zero when $\alpha > 0$. If the asymptotic zero density (or the asymptotic density of the eigenvalues of the random matrix) has a soft edge at the origin, then for $\alpha = 0$ one can use the Airy function for the local asymptotics at the origin, but for $\alpha \neq 0$ one needs Painlevé transcendents. They showed that for

$$\lim_{n,N\to\infty} n^{2/3} \left(\frac{n}{N} - 1 \right) = L,$$

one has

$$\lim_{n,N\to\infty} \frac{1}{(cn)^{2/3}} \widetilde{K}_{n,N} \left(\frac{u}{(cn)^{2/3}}, \frac{v}{(cn)^{2/3}} \right) = K_{\text{XXXIV}}(u, v; s),$$

where c and s are specific constants and the limiting kernel is given by

$$K_{\text{XXXIV}}(u, v; s) = \frac{\psi_2(u; s)\psi_1(v; s) - \psi_1(u; s)\psi_2(y; s)}{2\pi i(u - v)}$$

with ψ_1, ψ_2 special solutions of a model Riemann–Hilbert problem with prescribed asymptotic behavior near $\pm\infty$ on the real line. The model Riemann–Hilbert problem is related to a special solution of Painlevé XXXIV

$$u'' = 4u^2 + 2su + \frac{(u')^2 - (2\alpha)^2}{2u}, \tag{7.16}$$

which is named so because it appears as number XXXIV in the book of Ince [88, Ch. 14]. This differential equation can be transformed to Painlevé II.

Xu and Zhao [155] investigated the Hermite weight with a jump: they considered

$$w(x) = e^{-x^2} \begin{cases} 1, & x < \mu_n, \\ \omega, & x > \mu_n, \end{cases}$$

where ω is a complex number and μ_n is such that $\mu_n / \sqrt{2n} \to 1$ as $n \to \infty$. The asymptotic distribution of the scaled zeros $x_{j,n} / \sqrt{2n}$ of the corresponding orthogonal polynomials is on $[-1, 1]$ and the jump position of the jump after scaling is $\mu_n / \sqrt{2n}$ and tends toward the soft edge 1. So this situation is again a soft edge that meets a singularity of the weight. Again Painlevé XXXIV was used to construct the local parametrix near this soft edge. Xu and Zhao [155, Thm. 2] found the asymptotic behavior of the recurrence coefficients of the orthogonal polynomials in terms of a Painlevé XXXIV transcendent by employing the Riemann–Hilbert approach.

Exercise 16: Show that Painlevé XXXIV in (7.16) can be transformed to Painlevé II in (1.19) (with $\alpha \mapsto -2\alpha - \frac{1}{2}$) by using the transformation

$$u(s) = 2^{-1/3}U(-2^{1/3}s), \quad U(x) = y^2(x) + y'(x) + x/2,$$

and the inverse transformation

$$y(x) = -2^{1/3}Q(-2^{-1/3}x), \quad Q(s) = \frac{u'(s) - 2\alpha}{2u(s)}.$$

7.3 Painlevé III

Painlevé III asymptotics for orthogonal polynomials appears when the weight function of the orthogonality measure has an essential singularity. Brightmore, Mezzadri and Mo [20] investigated the singularly perturbed Hermite weight function

$$w(x, t) = \exp\left(-\frac{1}{2}x^2 - \frac{t}{2x^2}\right), \qquad x \in \mathbb{R},$$

with parameter $t > 0$ and found Painlevé III asymptotics for the orthogonal polynomials near 0 (actually, for the Christoffel–Darboux kernel near 0). Xu, Dai and Zhao [153] [154] used a singularly perturbed Laguerre weight

$$w(x, t) = x^\alpha \exp\left(-x - \frac{t}{x}\right), \qquad x > 0$$

with $\alpha > 0$ and $t > 0$ and also found Painlevé III asymptotics near the origin. What is typically happening for these weights is that the asymptotic density of the zeros (the density of the equilibrium measure) for small t lives on an interval $[-a_t, a_t]$ for the Hermite case, or on $[0, a_t]$ for the Laguerre case, but for large $t > 0$ the density will be on two disjoint intervals $[-a_t, -b_t] \cup [b_t, a_t]$ for the Hermite case, or an interval $[b_t, a_t]$, with $b_t > 0$ in the Laguerre case. Therefore, when $t \to 0$ there is a transition from two intervals to one (Hermite case) or from a soft edge $b_t > 0$ to a hard edge at the origin (Laguerre). We will not work out the details for this Painlevé III case, but refer to the papers mentioned above for the Riemann–Hilbert analysis and the final result.

Atkin, Claeys and Mezzadri [4] later analyzed singular perturbations of the Laguerre weight of the form

$$w(x, t) = x^\alpha \exp\left(-x - \frac{t^k}{x^k}\right), \qquad x > 0,$$

which for $k = 1$ gives asymptotic behavior in terms of Painlevé III, as described above, and for $k > 1$ gives asymptotics in terms of solutions in the hierarchy of the P_{III} equations.

7.4 Painlevé IV

When the density of the zeros vanishes linearly at a point, then Painlevé IV transcendents can be used for the local asymptotic behavior around that point. This cannot happen for orthogonal polynomials on the real line, but there are families of orthogonal polynomials in the complex plane for which this happens. Laguerre polynomials with a large negative parameter α are an example of this. For the Laguerre polynomials $L_n^{-N+\nu}(z)$, the weight function would be $w(z) = z^{-N+\nu} e^{-z}$ but this weight is not integrable at $z = 0$ if $-N + \nu \le -1$. For $\nu = -1$ these polynomials are the Taylor polynomials for the exponential function. One can obtain an orthogonality relation by integrating over a curve Γ in the complex plane from $\infty - i\epsilon$ to $\infty + i\epsilon$ which goes around the positive real axis $[0, \infty)$, so that one gets orthogonality (non-Hermitian) for such Laguerre polynomials

$$\int_\Gamma L_n^{-N+\nu}(z) L_m^{-N+\nu}(z) z^{-N+\nu} e^{-z}\, dz = 0, \qquad m \ne n.$$

In 1924 Szegő [145] proved that the zeros of $L_N^{-N+\nu}(Nz)$ accumulate on a curve S given by

$$S = \{|ze^{1-z}| = 1\} \cap \{|z \le 1\},$$

and this curve is known as the Szegő curve, see Figure 7.12. He proved this for $\nu = -1$ but the result holds for every $\nu \in \mathbb{R}$.

The asymptotic density of these zeros on the curve is known (see, e.g., [139]) and vanishes as $|z - 1|$ at the point $z = 1$. The local asymptotics around this singular point $z = 1$ can be obtained in terms of parabolic cylinder functions D_ν. Dai and Kuijlaars [48] considered an extension of this weight (and incorporated the scaling) by taking

$$w_N(z) = z^{-N+\nu} e^{-Nz} (z - 1)^{2b}.$$

The extra factor $(z - 1)^{2b}$ has influence on the local asymptotics near $z = 1$ and

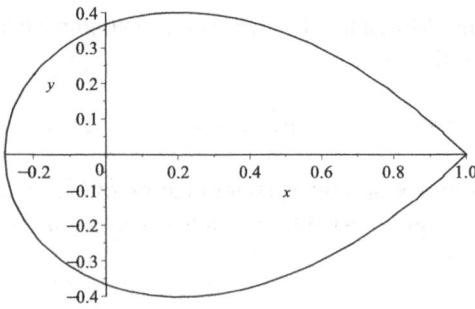

Figure 7.12 The Szegő curve $S = \{|ze^{1-z}| = 1\}$ inside the unit disk $|z| \leq 1$

the parabolic cylinder functions will no longer work when $b \neq 0$. One needs to use Painlevé transcendents. Recall that Painlevé IV has special function solutions in terms of parabolic cylinder functions (see Section 6.2.3), so a good guess is that Painlevé IV might work, and this is indeed true. We briefly show why and only consider the case $\nu > 0$ and not an integer.

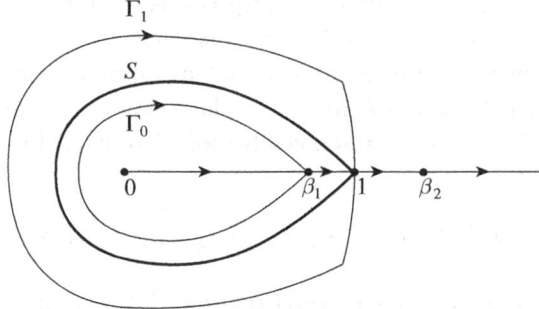

Figure 7.13 The contours for the Riemann–Hilbert problems for U, T and V

The Riemann–Hilbert problem for the corresponding orthogonal polynomials is on a curve Γ in the complex plane encircling the positive real axis $[0, \infty)$. For Laguerre polynomials with a negative parameter α it is known (see, e.g., [103]) that when $-\alpha/N \to A$, the (scaled) zeros accumulate on a closed curve and an interval if $A < 1$ and on the Szegő curve S when $A = 1$. In the case under consideration $\alpha = -N + \nu$, so $-\alpha/N = 1 - \nu/N$, and therefore the zero density will be supported on a curve Γ_0 and an interval $[\beta_1, \beta_2]$, with $\beta_1 < 1 < \beta_2$. As $N \to \infty$ the curve Γ_0 converges to the Szegő curve S and β_1, β_2 will converge to 1. We can deform the original contour Γ to the contour $\Gamma_0 \cup [\beta, \infty)$ and use the Riemann–Hilbert formulation for a matrix valued function U, with a jump on

Γ_0 and a jump on (β_1, ∞) and some control on what happens at $z = 1$. The next step is to normalize the Riemann–Hilbert problem so that the solution tends to the identity matrix as $z \to \infty$. For this one uses the logarithmic potential of the asymptotic zero distribution (the g-function). The new matrix valued function T will have an oscillatory jump on Γ_0 and on (β_1, β_2), and a jump that tends to the identity matrix on (β_2, ∞). To handle the oscillatory jumps, we open a lens Γ_1 around Γ_0 and we make sure that this curve Γ_1 is outside the Szegő curve S, see Figure 7.13. The new Riemann–Hilbert matrix V will have no jump on Γ_0 but has two new jumps on Γ_1 and on $(0, \beta_1)$. All the jumps in the region $|ze^{1-z}| < 1$ converge (pointwise) to the identity matrix, and the jump on Γ_1 also converges to the identity matrix. Near $z = 1$ the convergence is not uniform, so one needs to make a local analysis around $z = 1$ by constructing a local parametrix there. Near $z = 1$ the Riemann–Hilbert problem looks like Figure 7.14.

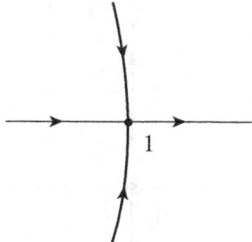

Figure 7.14 Parametrix around $z = 1$

This parametrix can be obtained by using the Riemann–Hilbert problem for Painlevé IV [68]. Consider the following pair of partial differential equations for the matrix function $\Psi(\lambda, s)$ (this is a *Lax pair* for P_{IV})

$$\frac{\partial \Psi}{\partial \lambda} = A\Psi, \qquad \frac{\partial \Psi}{\partial s} = B\Psi,$$

where

$$A = \left(\lambda + s + \frac{\Theta - K}{\lambda}\right)\sigma_3 + y\left(1 - \frac{u}{2\lambda}\right)\sigma_+ + \frac{2}{y}\left(K - \Theta - \Theta_\infty + \frac{K}{\lambda u}(K - 2\Theta)\right)\sigma_-,$$

$$B = \lambda\sigma_3 + y\sigma_+ + \frac{2}{y}(K - \Theta - \Theta_\infty)\sigma_-,$$

where the matrices $\sigma_3, \sigma_\pm$ are given by

$$\sigma_3 = \begin{pmatrix} 1 & 0 \\ 0 & -1 \end{pmatrix}, \quad \sigma_+ = \begin{pmatrix} 0 & 1 \\ 0 & 0 \end{pmatrix}, \quad \sigma_- = \begin{pmatrix} 0 & 0 \\ 1 & 0 \end{pmatrix},$$

and $y = y(s)$ and $K = K(s)$ are defined by

$$\frac{y'(s)}{y(s)} = -u - 2s, \quad K(s) = \frac{1}{4}(-u' + u^2 + 2su + 4\Theta). \qquad (7.17)$$

Then the compatibility between these two partial differential equations, i.e., the fact that

$$\frac{\partial^2 \Psi}{\partial \lambda \partial s} = \frac{\partial^2 \Psi}{\partial s \partial \lambda},$$

gives a condition on u, which is precisely the Painlevé IV equation (1.21) with $\alpha = 2\Theta_\infty - 1$ and $\beta = -8\Theta^2$. The function $\Psi(\lambda, s)$ satisfies a Riemann–Hilbert problem in the variable λ, with s a parameter in the asymptotic condition. The Riemann–Hilbert problem has jumps on two lines through the origin, with jumps as in Figure 7.15.

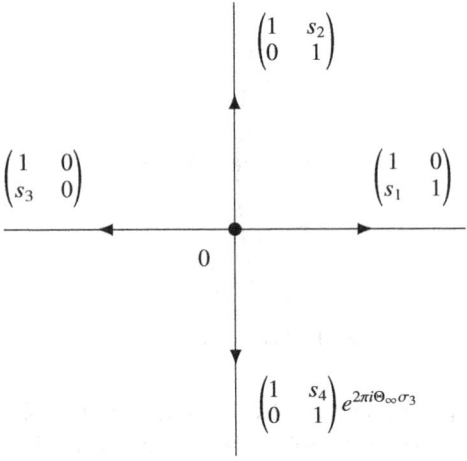

Figure 7.15 The Riemann–Hilbert problem for P$_{\mathrm{IV}}$

The numbers (s_1, s_2, s_3, s_4) are known as *Stokes multipliers* or Stokes parameters and they satisfy

$$(1 + s_2 s_3)e^{2\pi\Theta_\infty} + [s_1 s_4 + (1 + s_3 s_4)(1 + s_1 s_2)]e^{-2\pi i\Theta_\infty} = 2\cos 2\pi\Theta.$$

The asymptotic condition for the Riemann–Hilbert problem is for $\lambda \to \infty$

$$\Psi(\lambda, s) = \left(\mathbb{I} + \frac{\Psi_{-1}(s)}{\lambda} + \frac{\Psi_{-2}(s)}{\lambda^2} + O(1/\lambda^3)\right)e^{(\lambda^2/2 + s\lambda)\sigma_3}\lambda^{-\Theta_\infty\sigma_3},$$

and for some connection matrix C

$$\Psi(\lambda, s) = C^{-1}\lambda^{-\Theta\sigma_3} = O(1), \qquad \lambda \to 0, \Re\lambda > 0, \Im\lambda > 0.$$

A solution of Painlevé IV is then given by

$$u(s) = -2s = \frac{d}{ds} \log\left(\Psi_{-1}(s)\right)_{1,2}.$$

This solution will depend on the Stokes multipliers (s_1, s_2, s_3, s_4). If we compare the Riemann–Hilbert problem for the local parametrix with the Riemann–Hilbert problem for Painlevé IV, then the contours already agree but we need to shift the singularity 1 to 0. Furthermore we still need to check the jumps and the behavior on the boundary of the neighborhood in which we need the local parametrix. It turns out that one needs the Stokes multipliers

$$s_1 = \frac{\sin(\nu + 2b)\pi}{\sin \nu\pi} e^{(\nu+b)\pi i}, \quad s_2 = 2ie^{b\pi i} \sin \nu\pi,$$

$$s_3 = -e^{-(\nu+b)\pi i}, \qquad\qquad s_4 = -2ie^{(2\nu+b)\pi i} \sin \nu\pi,$$

and $\Theta = -b$, $\Theta_\infty = \nu + b$. The final result in [48] is the asymptotic behavior of the recurrence coefficients for the orthogonal polynomials:

Theorem 7.3 (Dai and Kuijlaars)   *Suppose $b \in \mathbb{R}$ and $\nu \in \mathbb{R} \setminus \mathbb{N}_0$ and put $\Theta = -b$ and $\Theta_\infty = \nu + b$. Let $u(s)$ be the solution of P_{IV} in (1.21) with $\alpha = 2\Theta_\infty - 1$ and $\beta = -8\Theta^2$ that corresponds to the Stokes parameters $(s_1/d, ds_2, s_3/d, ds_4)$, where $d = 2\pi i \sin \pi\nu$ and let $K(s)$ be as in (7.17). Assume that $\alpha = -n + \nu$ and $N = n + \sqrt{2nL}$, where L is not a pole of u. Then, for every large enough n, the monic orthogonal polynomial P_n of degree n for $w(z) = z^\alpha e^{-Nz}(z-1)^{2b}$ exists uniquely. The recurrence coefficients from (1.6) satisfy for $n \to \infty$*

$$a_n^2 = \frac{\nu - K(L)}{n} + O(n^{-3/2}),$$

and if $K(L) \neq \nu$

$$b_n = 1 + \sqrt{\frac{2}{n}} \left(\frac{K(L)(K(L) + 2b)}{u(L)(K(L) - \nu)} - L \right) + O(1/n).$$

7.5 Painlevé V

Painlevé V transcendents make their appearance when the weight function w depends on a parameter t and a singularity appears when $t \to t_*$, or when two existing singularities of w are merging when $t \to t_*$, where t_* is the critical parameter where the phase transition occurs. This has been studied for orthogonal polynomials on the unit circle, and the type of singularity that has been investigated in detail is a Fisher–Hartwig singularity [67]. Claeys, Its and Krasovsky

[30] have investigated the weight function

$$w(\theta, t) = |z - e^t|^{\alpha+\beta}|z - e^{-t}|^{\alpha-\beta} z^{-\alpha+\beta} e^{i\pi(\alpha+\beta)} e^{V(z)}, \qquad z = e^{i\theta},$$

where $t \geq 0$, $\Re\alpha > -\frac{1}{2}$, $\alpha \pm \beta \neq -1, -2, -3, \ldots$, and V is analytic in an annulus that contains the unit circle. For $t = 0$ the weight has a Fisher–Hartwig singularity, which is the combination of a jump singularity and an algebraic singularity at $z = 1$ (or $\theta = 0$). The asymptotic behavior of the Toeplitz determinant containing the trigonometric moments of this weight is well understood nowadays, see [62], [52]. For $t > 0$ the weight has two algebraic singularities at $z = e^t$ and $z = e^{-t}$ and these are not on the unit circle, so that the asymptotic behavior of the Toeplitz determinant follows from Szegő's result for Toeplitz determinants. In [30] it is shown that one needs a solution of the Jimbo–Miwa σ-form of Painlevé V, i.e., the differential equation

$$(x\sigma'')^2 = (\sigma - x\sigma' + 2(\sigma')^2 + 2\alpha\sigma')^2 - 4(\sigma')^2(\sigma' + \alpha + \beta)(\sigma' + \alpha - \beta),$$

to describe the asymptotic behavior of the Toeplitz determinant for small t. In a later paper, Claeys and Krasovsky [31] investigated the situation when two Fisher–Hartwig singularities at e^{it} and e^{-it} merge as $t \to 0$. They also need the solution of a the σ-form of Painlevé V. We will not explain their results here, but refer to their papers for the details.

Instead, we will explain a similar situation for orthogonal polynomials on the real line, which was studied by Claeys and Fahs [29]. They investigated a random matrix model for $n \times n$ Hermitian matrices M with probability density

$$\frac{1}{Z_n} |\det(M^2 - tI)|^\alpha e^{-n\operatorname{Tr} V(M)} \, dM,$$

with $\alpha > -\frac{1}{2}$ and $t \in \mathbb{R}$. The eigenvalues of such random matrices can be studied by using the orthogonal polynomials on the real line for the weight function

$$w(x, t) = |x^2 - t|^\alpha e^{-nV(x)}, \qquad x \in \mathbb{R}. \tag{7.18}$$

It is assumed that V is such that the equilibrium distribution is supported on one interval $[a, b]$ and that $a < 0 < b$. The equilibrium distribution has a density $\rho(x) = \sqrt{(x-a)(b-x)} h(x)$, with h a positive function on $[a, b]$, independent of α and t. If $t = 0$ then the weight (7.18) has an algebraic singularity at 0 and the zeros/eigenvalues behave as though they are distributed on two touching intervals $(a, 0]$ and $[0, b)$, with 0 behaving as a hard edge and a, b behaving as soft edges. When $t > 0$ the weight (7.18) has two singularities at $x = \pm\sqrt{t}$ and for $t \to 0$ these two singularities are merging to one singularity. For $t < 0$ there are no singularities on the real line, but the two imaginary singularities at $\pm i \sqrt{|t|}$ move to 0 as $t \to 0$ so that a singularity appears when $t \to 0$. Both situations were analyzed in [29], but we only deal with $t > 0$. Here is the result when $n \to \infty$ and $t \to 0$, with $t \sim 1/n^2$.

Theorem 7.4 (Claeys–Fahs) *Let $\tau_{n,t} = 16\pi^2\rho(0)^2 n^2 t$ and let $n \to \infty$ and $t \to 0$.*

1. *If $\tau_{n,t} \to \pm\infty$, then*

$$\lim_{n\to\infty} \frac{1}{n\rho(0)} \widetilde{K}_n\left(\frac{u}{n\rho(0)}, \frac{v}{n\rho(0)}\right) = K_{\sin}(\pi u, \pi v),$$

   where $K_{\sin}$ is given in (7.3).
2. *If $\tau_{n,t} \to 0$ and $u, v \neq 0$, then*

$$\lim_{n\to\infty} \frac{1}{n\rho(0)} \widetilde{K}_n\left(\frac{u}{n\rho(0)}, \frac{v}{n\rho(0)}\right) = K_{\text{Bessel}}^{\alpha+\frac{1}{2}}(\pi u, \pi v),$$

   where K_{Bessel} is given in (7.5).
3. *If $\tau_{n,t} \to \tau \neq 0$, and $u, v \neq \pm \sqrt{\tau}/4$, then*

$$\frac{1}{n\rho(0)} \widetilde{K}_n\left(\frac{u}{n\rho(0)}, \frac{v}{n\rho(0)}\right) = K_V^\tau(\pi u, \pi v),$$

   where

$$K_V^\tau(s, t) = \frac{\Phi_1(v; \tau)\Phi_2(u; \tau) - \Phi_1(u; \tau)\Phi_2(v; \tau)}{2\pi i(u - v)},$$

   where Φ_1 and Φ_2 are special solutions of a Lax pair which is related to the fifth Painlevé equation.

The Lax pair which is mentioned in the above theorem is

$$\frac{\partial}{\partial z}\Psi(z; s) = A(z; s)\Psi(z; s),$$

$$\frac{\partial}{\partial s}\Psi(z; s) = B(z; s)\Psi(z; s),$$

with

$$A(z; s) = \frac{-s}{2}\begin{pmatrix} 1 & 0 \\ 0 & -1 \end{pmatrix} + \frac{1}{z}\begin{pmatrix} -v + \alpha/2 & uy(v - \alpha) \\ -v/uy & v - \alpha/2 \end{pmatrix} + \frac{1}{z-1}\begin{pmatrix} v - \alpha/2 & -y(v - \alpha) \\ v/y & -v + \alpha/2 \end{pmatrix},$$

and

$$B(z; s) = \frac{1}{s}\begin{pmatrix} 0 & y(v - \alpha)(u - 1) \\ \frac{v(u-1)}{yu} & 0 \end{pmatrix} - \frac{z}{2}\begin{pmatrix} 1 & 0 \\ 0 & -1 \end{pmatrix}.$$

The functions $u(s)$ and $v(s)$ are related to the functions p, q, r by

$$v = \frac{\alpha}{2} - q - srp, \quad u = 1 + \frac{sp}{(1 - s)p + sp'}.$$

The compatibility

$$\frac{\partial^2}{\partial s \partial t}\Psi(z; s) = \frac{\partial^2}{\partial t \partial s}\Psi(z; s)$$

gives the second order differential equation for $u = u(s)$

$$u'' = \left(\frac{1}{2u} + \frac{1}{u-1}\right)(u')^2 - \frac{u'}{s} + \frac{(1-u)^2}{s^2}\frac{\alpha^2}{2}\left(u - \frac{1}{u}\right) + \frac{u}{s} - \frac{u(u+1)}{2(u-1)},$$

which is a special case of P_V in (1.22). The functions Φ_1 and Φ_2 are then given by

$$\begin{pmatrix}\Phi_1(x; \tau)\\ \Phi_2(x; \tau)\end{pmatrix} = \Psi(z; s)\Delta(z), \qquad z = \frac{2x}{|s|} + \frac{1}{2},$$

where

$$\Delta(z) = \begin{cases} \begin{pmatrix} e^{-i\pi\alpha/2}\\ -e^{i\pi\alpha/2}\end{pmatrix}, & \text{if } \Re z > 1,\\[12pt] \begin{pmatrix} 0\\ -1\end{pmatrix}, & \text{if } 0 < \Re z < 1,\\[12pt] \begin{pmatrix} e^{i\pi\alpha/2}\\ -e^{-i\pi\alpha/2}\end{pmatrix}, & \text{if } \Re z < 0.\end{cases}$$

The matrix Ψ solves a Riemann–Hilbert problem on a system of contours as is given in Figure 7.16 with behavior for $z \to \infty$ given by

$$\Psi(z; s) = \left(\mathbb{I} + \frac{1}{z}\begin{pmatrix} q & r\\ p & -q\end{pmatrix} + O(\frac{1}{z^2})\right)\begin{pmatrix} e^{-sz/2} & 0\\ 0 & e^{sz/2}\end{pmatrix}.$$

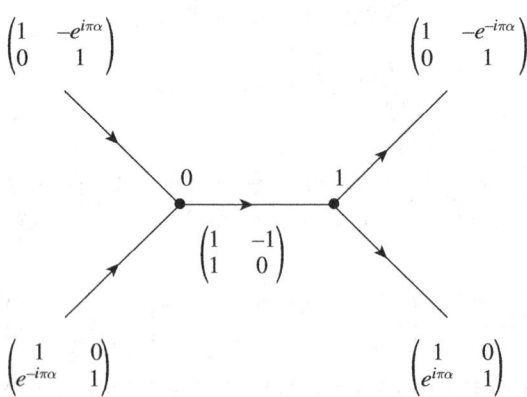

Figure 7.16 The Riemann–Hilbert problem for P_V

To indicate how Painlevé V enters into the asymptotic analysis of orthogonal polynomials with merging singularities, one uses the Riemann–Hilbert problem for orthogonal polynomials with the weight (7.18). The original contour for the jump condition is the full real line $\mathbb{R}$, but after normalizing the Riemann–Hilbert problem for $z \to \infty$, which uses the equilibrium measure with density ρ on $[a, b]$, the Riemann–Hilbert problem has jumps close to the identity matrix on $(-\infty, a)$ and (b, ∞), and an oscillatory jump on (a, b). The oscillatory jump can be handled by the Deift–Zhou steepest descent method. However, for opening a lens around (a, b), one only opens a lens around $(a, -\sqrt{t})$ and $(\sqrt{t}, b)$, where $\pm\sqrt{t}$ are the algebraic singularities in the weight (7.18), see Figure 7.17. Recall that $t = O(1/n^2)$ in case 3 of Theorem 7.4, so that $\pm\sqrt{t} = O(1/n)$.

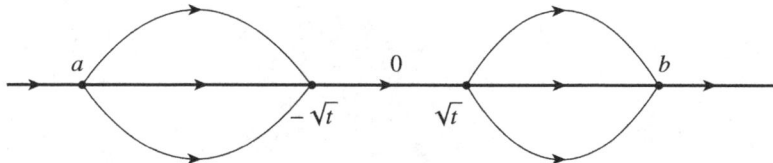

Figure 7.17 Opening the lenses for merging singularities

Then one neglects all the jumps that tend to the identity matrix and solves the resulting Riemann–Hilbert problem, which gives the global parametrix containing the dominant terms in the asymptotic behavior. The jumps that converge to the identity matrix, however, do not converge uniformly in the neighborhood of a, b and 0, so one needs to construct local parametrices around these points. Around a and b this local parametrix can be constructed using Airy functions. Around 0 one needs a new construction. Observe that any neighborhood of the origin will contain the points $\pm\sqrt{t}$ when n is large enough. Inside a neighborhood of the origin one uses the parametrix

$$P(z) = E(z)\Psi(\lambda(z); s_{n,t})W(z)$$

where $E(z)$ is such that it matches the global parametrix on the boundary of the neighborhood, $s_{n,t} = -i\sqrt{\tau_{n,t}} + O(nt^{3/2})$, λ is a conformal map in the neighborhood of the origin and W is related to the weight (7.18). Observe that the contours inside a neighborhood of the origin indeed look like the contours of the Riemann–Hilbert problem for Ψ in Figure 7.16.

7.6 Painlevé VI

As far as I know there is nothing in the literature about using the Painlevé VI transcendents for constructing a local parametrix at certain critical points in the asymptotic behavior of orthogonal polynomials or eigenvalues of random matrices. It is an open problem to find a situation where such Painlevé transcendents are needed.

Appendix
Solutions to the exercises

Exercise 1

The weight function for Laguerre polynomials is $w(x) = x^\alpha e^{-x}$ on $[0, \infty)$, where $\alpha > -1$. The Pearson equation is

$$[xw(x)]' = (\alpha + 1 - x)w(x). \tag{A.1}$$

The structure relation is therefore of the form

$$xp_n'(x) = A_n p_n(x) + B_n p_{n-1}, \qquad n \geq 1.$$

Taking derivatives in the three term recurrence relation gives

$$p_n(x) + xp_n'(x) = a_{n+1}p_{n+1}'(x) + b_n p_n'(x) + a_n p_{n-1}'(x).$$

Multiply this by x and use the structure relation, to find

$$xp_n(x) + x(A_n p_n(x) + B_n p_{n-1}(x)) = a_{n+1}(A_{n+1}p_{n+1}(x) + B_{n+1}p_n(x))$$
$$+ b_n(A_n p_n(x) + B_n p_{n-1}(x)) + a_n(A_{n-1}p_{n-1}(x) + B_{n-1}p_{n-2}(x)).$$

Finally, use the three term recurrence relation on the left, to arrive at

$$(A_n + 1)(a_{n+1}p_{n+1}(x) + b_n p_n(x) + a_n p_{n-1}(x))$$
$$+ B_n(a_n p_n(x) + b_{n-1}p_{n-1}(x) + a_{n-1}p_{n-2}(x)) = a_{n+1}(A_{n+1}p_{n+1}(x) + B_{n+1}p_n(x))$$
$$+ b_n(A_n p_n(x) + B_n p_{n-1}(x)) + a_n(A_{n-1}p_{n-1}(x) + B_{n-1}p_{n-2}(x)).$$

Equating coefficients in front of the polynomials p_{n+1}, p_n, p_{n-1} and p_{n-2} gives

$$p_{n+1} \Rightarrow a_{n+1}(A_n + 1) = a_{n+1}A_{n+1}, \tag{A.2}$$
$$p_n \Rightarrow (A_n + 1)b_n + B_n a_n = a_{n+1}B_{n+1} + b_n A_n, \tag{A.3}$$
$$p_{n-1} \Rightarrow (A_n + 1)a_n + B_n b_{n-1} = b_n B_n + a_n A_{n-1}, \tag{A.4}$$
$$p_{n-2} \Rightarrow B_n a_{n-1} = a_n B_{n-1}. \tag{A.5}$$

147

From (A.2) we find that $A_{n+1} - A_n = 1$ so that $A_n = n$. From (A.5) we find that $B_n/a_n = B_{n-1}/a_{n-1}$ so that this ratio is constant and $B_n = ca_n$, with c a constant. From (A.3) we then get

$$b_n = c(a_{n+1}^2 - a_n^2),$$

and from (A.4)

$$c(b_n - b_{n-1}) = 2.$$

Summing this from 1 to n gives $b_n = 2n/c + b_0$, and if we use this in the previous equation, then summing from 1 to $n-1$ gives $a_n^2 = n(n-1)/c + nb_0$. It remains to determine the constants b_0 and c. For b_0 we use the general formula $b_0 = m_1/m_0$, where m_0 and m_1 are the first two moments,

$$m_k = \int_0^\infty x^k x^\alpha e^{-x}\, dx = \Gamma(k + \alpha + 1).$$

Hence $b_0 = \alpha + 1$, where we used $\Gamma(\alpha + 2) = (\alpha + 1)\Gamma(\alpha + 1)$. For c we compute B_1 in the structure relation:

$$B_1 = \int_0^\infty xp_1'(x)p_0 w(x)\, dx = -\int_0^\infty p_1(x)p_0[xw(x)]'\, dx$$

and with the Pearson equation (A.1) we get

$$B_1 = -\int_0^\infty p_1(x)p_0(x)(\alpha + 1 - x)w(x)\, dx = a_1,$$

where we used (1.4). Hence $c = 1$ and with this information we find

$$a_n^2 = n(n + \alpha), \qquad b_n = 2n + \alpha + 1.$$

Exercise 2

If we put $\alpha = 2h^5, \beta = -3, a = 0$ and $b = 6$ in (1.24), then

$$x_n(x_{n+1} + x_n + x_{n-1}) = 2h^5 n - 3 + 6x_n.$$

Now fix x and h so that $x = nh$, then $x + h = (n + 1)h$ and $x - h = (n - 1)h$, so that we replace n on the right-hand side by x/h and set $x_n = 1 - 2h^2 y(x)$, $x_{n+1} = 1 - 2h^2 y(x + h)$ and $x_{n-1} = 1 - 2h^2 y(x - h)$. This gives

$$(1 - 2h^2 y(x))(3 - 2h^2[y(x + h) + y(x) + y(x - h)]) = 2xh^4 - 3 + 6[1 - 2h^2 y(x)].$$

Working this out and collecting powers of h gives

$$-2h^2\big(y(x + h) - 2y(x) + y(x - h)\big) + 4h^4 y(x)\big(y(x + h) + y(x) + y(x - h)\big) = 2xh^4.$$

Now divide by h^4 and let h tend to zero, and use

$$\lim_{h \to 0} \frac{y(x+h) - 2y(x) + y(x-h)}{h^2} = y''(x),$$

then

$$-2y''(x) + 12y^2(x) = 2x,$$

which is the same $y'' = 6y^2 - x$ and this is the equation in (1.18) but with $-x$ instead of x. If we put $y(x) = u(-x)$, then we have $u''(-x) = 6u^2(-x) - x$, so that $x = -t$ gives Painlevé I with the correct sign.

A possible choice of the parameters in discrete Painlevé II is to take $\alpha = h^3$, $\beta = 2$, $a = \gamma h^3$, then (1.25) becomes

$$(1 - x_n^2)(x_{n+1} + x_{n-1}) = x_n(h^3 n + 2) + \gamma h^3.$$

Fix x and h so that $x = nh$, then $x + h = (n+1)h$ and $x - h = (n-1)h$. So we can replace n on the right-hand side by x/h and put $x_n = hy(x)$, $x_{n+1} = hy(x+h)$ and $x_{n-1} = hy(x-h)$. This gives

$$h(1 - h^2 y^2(x))(y(x+h) + y(x-h)) = hy(x)(xh^2 + 2) + \gamma h^3.$$

If we work this out and collect powers of h, then

$$h\big(y(x+h) - 2y(x) + y(x-h)\big) - h^3 y^2(x)\big(y(x+h) + y(x-h)\big) = h^3 xy(x) + \gamma h^3.$$

Divide by h^3 and let $h \to 0$, then this gives

$$y''(x) - 2y^3(x) = xy(x) + \gamma,$$

which is Painlevé II in (1.19) with $\alpha = \gamma$.

Exercise 3

The weight function $w(x) = |x|^\rho e^{-x^4}$ satisfies the Pearson equation

$$[xw(w)]' = (\rho + 1 - 4x^4)w(x), \tag{A.6}$$

for $x > 0$ and for $x < 0$ (this can be checked for both cases separately). The structure relation is therefore of the form

$$xp_n'(x) = A_n p_n(x) + B_n p_{n-1}(x) + C_n p_{n-2}(x) + D_n p_{n-3}(x) + E_n p_{n-4}(x),$$

but since w is a symmetric weight function on $\mathbb{R}$, the formula simplifies to

$$xp_n'(x) = A_n p_n(x) + C_n p_{n-2}(x) + E_n p_{n-4}(x). \tag{A.7}$$

The three term recurrence relation is of the form

$$xp_n(x) = a_{n+1}p_{n+1}(x) + a_n p_{n-1}(x).$$

Take derivatives to find

$$p_n(x) + xp_n'(x) = a_{n+1}p_{n+1}'(x) + a_n p_{n-1}'(x),$$

then multiply by x and use the structure relation (A.7) to get

$$xp_n(x) + x\Big(A_n p_n(x) + C_n p_{n-2}(x) + E_n p_{n-4}(x)\Big)$$
$$= a_{n+1}\Big(A_{n+1}p_{n+1}(x) + C_{n+1}p_{n-1}(x) + E_{n+1}p_{n-2}(x)\Big)$$
$$+ a_n\Big(A_{n-1}p_{n-1}(x) + C_{n-1}p_{n-3}(x) + E_{n-1}p_{n-5}(x)\Big).$$

Now use the three term recurrence relation to find

$$(A_n + 1)\Big(a_{n+1}p_{n+1}(x) + a_n p_{n-1}(x)\Big)$$
$$+ C_n\Big(a_{n-1}p_{n-1}(x) + a_{n-2}p_{n-3}(x)\Big) + E_n\Big(a_{n-3}p_{n-3}(x) + a_{n-4}p_{n-5}(x)\Big)$$
$$= a_{n+1}\Big(A_{n+1}p_{n+1}(x) + C_{n+1}p_{n-1}(x) + E_{n+1}p_{n-2}(x)\Big)$$
$$+ a_n\Big(A_{n-1}p_{n-1}(x) + C_{n-1}p_{n-3}(x) + E_{n-1}p_{n-5}(x)\Big).$$

Equating coefficients in front of the polynomials $p_{n+1}, p_{n-1}, p_{n-3}$ and p_{n-5} then gives

$$p_{n+1} \Rightarrow (A_n + 1)a_{n+1} = a_{n+1}A_{n+1}, \tag{A.8}$$
$$p_{n-1} \Rightarrow (A_n + 1)a_n + C_n a_{n-1} = a_{n+1}C_{n+1} + a_n A_{n-1}, \tag{A.9}$$
$$p_{n-3} \Rightarrow C_n a_{n-2} + E_n a_{n-3} = a_{n+1}E_{n+1} + a_n C_{n-1}, \tag{A.10}$$
$$p_{n-5} \Rightarrow E_n a_{n-4} = a_n E_{n-1}. \tag{A.11}$$

From (A.8) we get $A_{n+1} - A_n = 1$ so that $A_n = n$, and from (A.11) we see that $E_n / a_n a_{n-1}a_{n-2}a_{n-3} = E_{n-1}/a_{n-1}a_{n-2}a_{n-3}a_{n-4}$ so that this ratio is constant: $E_n = ca_n a_{n-1}a_{n-2}a_{n-3}$, with c a constant. We can find this constant by computing the first coefficient E_4, which is the coefficient of p_0 in the Fourier expansion of $xp_4'(x)$ in (A.7), hence

$$E_4 = \int_{-\infty}^{\infty} xp_4'(x)p_0 w(x)\, dx = -\int_{-\infty}^{\infty} p_4(x)p_0[xw(x)]'\, dx,$$

where we used integration by parts. The Pearson equation (A.6) gives

$$E_4 = -\int_{-\infty}^{\infty} p_4(x)p_0(\rho + 1 - 4x^4)w(x)\, dx = 4\int_{-\infty}^{\infty} x^4 p_4(x)p_0 w(x)\, dx.$$

Now use $x^4 = p_4(x)/\gamma_4 + \cdots$ to find

$$E_4 = 4\frac{p_0}{\gamma_4} = 4a_1 a_2 a_3 a_4,$$

where we used $p_0 = \gamma_0$ and (1.3). Hence $c = 4$ and $E_n = 4a_n a_{n-1} a_{n-2} a_{n-3}$. Use this in (A.10) to find

$$C_n a_{n-2} - a_n C_{n-1} = 4a_n a_{n-1} a_{n-2}(a_{n+1}^2 - a_{n-3}^2),$$

which, after dividing by the common factor on the right, is

$$\frac{C_n}{a_n a_{n-1}} - \frac{C_{n-1}}{a_{n-1} a_{n-2}} = 4(a_{n+1}^2 - a_{n-3}^2).$$

Both sides of this equation are differences, so we can sum easily from 3 to n to find

$$\frac{C_n}{a_n a_{n-1}} = \frac{C_2}{a_2 a_1} + 4(a_{n+1}^2 + a_n^2 + a_{n-1}^2 + a_{n-2}^2 - a_3^2 - a_2^2 - a_1^2). \tag{A.12}$$

Here we can determine C_2, which is the coefficient of p_0 in the Fourier expansion of $xp_2'(x)$ in (A.7), hence

$$C_2 = \int_{-\infty}^{\infty} xp_2'(x)p_0 w(x)\, dx = -\int_{-\infty}^{\infty} p_2(x) p_0 [xw(x)]'\, dx.$$

The Pearson equation (A.6) gives

$$C_2 = -\int_{-\infty}^{\infty} p_0 p_2(x)(\rho + 1 - 4x^4) w(x)\, dx = 4\int_{-\infty}^{\infty} x^4 p_0 p_2(x) w(x)\, dx.$$

This integral can be computed by using the three term recurrence relation four times or by using $p_2(x) = (x^2 - a_1^2)p_0/a_1 a_2$, and gives

$$C_2 = 4a_1 a_2(a_1^2 + a_2^2 + a_3^2),$$

so that (A.12) is

$$\frac{C_n}{a_n a_{n-1}} = 4(a_{n+1}^2 + a_n^2 + a_{n-1}^2 + a_{n-2}^2). \tag{A.13}$$

Finally, from (A.9) we find

$$2a_n^2 = a_n a_{n+1} C_{n+1} - a_{n-1} a_n C_n. \tag{A.14}$$

We now want to eliminate the sequence $(C_n)_{n \geq 2}$ from (A.13) and (A.14). To achieve this, we multiply (A.13) by $a_n^2 a_{n-1}^2$

$$a_n a_{n-1} C_n = 4a_n^2 a_{n-1}^2(a_{n+1}^2 + a_n^2 + a_{n-1}^2 + a_{n-2}^2),$$

and then insert this in (A.14) to find

$$2a_n^2 = 4a_n^2 a_{n+1}^2(a_{n+2}^2 + a_{n+1}^2 + a_n^2 + a_{n-1}^2) - 4a_n^2 a_{n-1}^2(a_{n+1}^2 + a_n^2 + a_{n-1}^2 + a_{n-2}^2).$$

Here $2a_n^2$ is a common factor and two terms on the right-hand side cancel, leaving

$$1 = 2a_{n+1}^2(a_{n+2}^2 + a_{n+1}^2 + a_n^2) - 2a_{n-1}^2(a_n^2 + a_{n-1}^2 + a_{n-2}^2).$$

If we write $c_n = 2a_n^2(a_{n+1}^2 + a_n^2 + a_{n-1}^2)$, then this simplifies to

$$1 = c_{n+1} - c_{n-1},$$

which contains a difference of step size 2. The best way is to consider even and odd n separately. So let $n = 2k$, then summing $1 = c_{2k+1} - c_{2k-1}$ from $k = 1$ to m gives

$$m = c_{2m+1} - c_1 = 2a_{2m+1}^2(a_{2m+2}^2 + a_{2m+1}^2 + a_{2m}^2) - c_1, \qquad (A.15)$$

and for $n = 2k - 1$ we sum $1 = c_{2k} - c_{2k-2}$ from $k = 1$ to m to find

$$m = c_{2m} - c_0 = 2a_{2m}^2(a_{2m+1}^2 + a_{2m}^2 + a_{2m-1}^2) - c_0. \qquad (A.16)$$

Now $c_0 = 0$ (since $a_0^2 = 0$) and $c_1 = 2a_1^2(a_2^2 + a_1^2)$. For the latter we use Parseval's identity

$$a_1^2 + a_2^2 = \int_{-\infty}^{\infty} [xp_1(x)]^2 w(x)\, dx,$$

and $p_1(x) = xp_0/a_1$ then gives

$$a_1^2 + a_2^2 = \frac{p_0^2}{a_1^2} m_4,$$

where m_4 is the fourth moment. Integrating the Pearson equation (A.6) gives $4m_4 = (\rho + 1)m_0$, hence $a_1^2(a_1^2 + a_2^2) = m_4/m_0 = (\rho + 1)/4$. So (A.15)–(A.16) indeed give the required recurrence

$$4a_n^2(a_{n+1}^2 + a_n^2 + a_{n-1}) = n + \begin{cases} 0, & \text{if } n \text{ is even,} \\ \rho, & \text{if } n \text{ is odd.} \end{cases}$$

Exercise 4

Let $x_n(t) = \frac{1}{2}y(-t/2)$, and denote by $'$ the derivative with respect to t, then

$$x_n' = -\frac{1}{4}y'(-t/2), \qquad x_n'' = \frac{1}{8}y''(-t/2).$$

Insert this in equation (2.18), then

$$y''(-t/2) = \frac{[y'(-t/2)]^2}{2y(-t/2)} + \frac{3[y(-t/2)]^3}{2} - 2[y(-t/2)]^2 t$$

$$+ 4y(-t/2)\left(\frac{n}{4} + \frac{t^2}{8}\right) - \frac{n^2}{2y(-t/2)}.$$

Now change t to $-2x$ to find

$$y''(x) = \frac{[y'(x)]^2}{2y(x)} + \frac{3[y(x)]^3}{2} + 4[y(x)]^2 + 2y(x)\left(\frac{n}{2} + x^2\right) - \frac{n^2}{2y(x)},$$

where now $'$ is the derivative with respect to x. This is the fourth Painlevé equation in (1.21) with parameters $\alpha = -n/2$, $\beta = -n^2/2$.

Exercise 5

Let

$$\alpha_n = \frac{1+y}{1-y},$$

then

$$\alpha_n' = \frac{2y'}{(1-y)^2}, \qquad \alpha_n'' = \frac{2y''(1-y) + 4(y')^2}{(1-y)^3},$$

and

$$1 - \alpha_n^2 = \frac{-4y}{(1-y)^2}.$$

Insert this in (3.13), then

$$\frac{2y''}{(1-y)^2} + \frac{4(y')^2}{(1-y)^3}$$

$$= \frac{(1+y)(y')^2}{y(1-y)^3} - \frac{2y'}{t(1-y)^2} + \frac{4(1+y)y}{(1-y)^3} - \frac{(n+1)^2(1+y)(1-y)}{4t^2y}.$$

Multiply this by $(1-y)^2$ and divide by 2 to get

$$y'' = -\frac{2(y')^2}{1-y} + \frac{(1+y)(y')^2}{2y(1-y)} - \frac{y'}{t} + \frac{2(1+y)y}{1-y} - \frac{(n+1)^2(1+y)(1-y)^3}{8t^2y}.$$

The terms containing $(y')^2$ are

$$(y')^2\left(\frac{-2}{1-y} + \frac{1+y}{2y(1-y)}\right) = (y')^2\left(\frac{1}{2y} + \frac{1}{y-1}\right).$$

The differential equation thus becomes

$$y'' = \left(\frac{1}{2y} + \frac{1}{y-1}\right)(y')^2 - \frac{y'}{t} + \frac{(n+1)^2}{8t^2}(y-1)^2\left(y - \frac{1}{y}\right) - \frac{2y(y+1)}{y-1},$$

which is the fifth Painlevé equation (1.22) with $\alpha = -\beta = (n+1)^2/8$, $\gamma = 0$ and $\delta = -2$.

Exercise 6

The expression $\sigma(x)\Delta p_n(x)$ is a polynomial of degree $n+s-1$, where $s = \deg \sigma$, hence we can expand it as a sum in terms of the orthonormal polynomials of degree $\leq n + s - 1$

$$\sigma\Delta p_n(x) = \sum_{j=0}^{n+s-1} A_{n,j}p_j(x).$$

The coefficients $A_{n,j}$ are Fourier coefficients in this expansion, and are thus given by

$$A_{n,j} = \sum_{k=0}^{\infty} \sigma(k)[\Delta p_n(k)]p_j(k)w_k.$$

Use summation by parts and $w_{-1} = w(-1) = 0$ to find

$$A_{n,j} = -\sum_{k=0}^{\infty} p_n(k)\nabla[\sigma(k)p_j(k)w_k].$$

In this sum we have

$$\begin{aligned}
\nabla[\sigma(k)p_j(k)w_k] &= \sigma(k)p_j(k)w_k - \sigma(k-1)p_j(k-1)w_{k-1} \\
&= \sigma(k)w_k[p_j(k) - p_j(k-1)] \\
&\quad + [\sigma(k)w_k - \sigma(k-1)w_{k-1}]p_j(k-1) \\
&= \sigma(k)w_k\nabla p_j(k) + p_j(k-1)\nabla[\sigma(k)w_k],
\end{aligned}$$

so that

$$A_{n,j} = -\sum_{k=0}^{\infty} p_n(k)[\nabla p_j(k)]\sigma(k)w_k - \sum_{k=0}^{\infty} p_n(k)p_j(k-1)\nabla[\sigma(k)w_k].$$

In the first sum the polynomial $\sigma(x)\nabla p_j(x)$ has degree $s + j - 1$, hence by orthogonality the sum will be zero whenever $j + s - 1 < n$. In the second sum we use the Pearson equation to find

$$\sum_{k=0}^{\infty} p_n(k)p_j(k-1)\nabla[\sigma(k)w_k] = \sum_{k=0}^{\infty} p_n(k)p_j(k-1)\tau(k)w_k,$$

and since $p_j(x-1)\tau(x)$ is a polynomial of degree $j+\deg\tau$, the sum will be zero whenever $j + \deg\tau < n$. So all the terms $A_{n,j}$ are zero whenever $j < n - s + 1$ and $j < n - \deg\tau$, hence for $j < n - t$, where $t = \max\{\deg\tau, s - 1\}$.

Exercise 7

If we put $x_n = iX_n/\sqrt{aB}$ and $iD = \sqrt{B/a}$, then the D_4^c equation becomes

$$-\frac{X_{n+1}X_n}{aB} = \frac{(y_n - z_n)^2 - A}{y_n^2 - B},$$

$$y_n + y_{n-1} = \frac{z_{n-1/2} - C}{1 + X_n/a} + \frac{z_{n-1/2} + C}{1 - X_n/B}.$$

Multiply the first equation by B and let $B \to \infty$, then we get

$$X_{n+1}X_n = a\big((y_n - z_n)^2 - A\big),$$

$$y_n + y_{n-1} = \frac{z_{n-1/2} - C}{1 + X_n/a} + z_{n-1/2} + C.$$

We claim that this is (3.38)–(3.39) for

$$X_n = a_n^2 - a, \quad y_n = b_n.$$

Complete squares to find

$$d_n(d_n + \beta - 1) = \left(d_n + \frac{\beta - 1}{2}\right)^2 - \left(\frac{\beta - 1}{2}\right)^2,$$

and recall that $d_n = b_n - n$, then we see that

$$z_n = n - \frac{\beta - 1}{2}, \quad A = \left(\frac{\beta - 1}{2}\right)^2.$$

If we use all this in the second equation, then

$$b_n + b_{n-1} = \frac{n - \beta/2 - C}{a_n^2/a} + n - \frac{\beta}{2} + C,$$

and from (3.39) we have

$$b_n + b_{n-1} = d_n + d_{n-1} + 2n - 1 = \frac{an}{a_n^2} + n - \beta,$$

hence this gives

$$C = -\frac{\beta}{2}.$$

Exercise 8

Start from

$$x\frac{d}{dt}p_n(x;t) = \left(\frac{d}{dt}a_{n+1}\right)p_{n+1}(x;t) + \left(\frac{d}{dt}b_n\right)p_n(x;t) + \left(\frac{d}{dt}a_n\right)p_{n-1}(x;t)$$

$$+ a_{n+1}\frac{d}{dt}p_{n+1}(x;t) + b_n\frac{d}{dt}p_n(x;t) + a_n\frac{d}{dt}p_{n-1}(x;t),$$

multiply this by $p_n(x;t)$ and then integrate with respect to the measure μ_t. Then

$$\int x\left(\frac{d}{dt}p_n(x;t)\right)p_n(x;t)\,d\mu_t(x) = \frac{d}{dt}b_n$$

$$+ a_{n+1}\int\left(\frac{d}{dt}p_{n+1}(x;t)\right)p_n(x;t)\,d\mu_t(x) + b_n\int\left(\frac{d}{dt}p_n(x;t)\right)p_n(x;t)\,d\mu_t(x),$$

where we used the orthonormality. If we use the three term recurrence relation on the left-hand side, then

$$\int x\left(\frac{d}{dt}p_n(x;t)\right)p_n(x;t)\,d\mu_t(x)$$

$$= b_n\int\left(\frac{d}{dt}p_n(x;t)\right)p_n(x;t)\,d\mu_t(x) + a_n\int\left(\frac{d}{dt}p_n(x;t)\right)p_{n-1}(x;t)\,d\mu_t(x),$$

so that

$$\frac{d}{dt}b_n = a_n\int\left(\frac{d}{dt}p_n(x;t)\right)p_{n-1}(x;t)\,d\mu_t(x)$$

$$- a_{n+1}\int\left(\frac{d}{dt}p_{n+1}(x;t)\right)p_n(x;t)\,d\mu_t(x).$$

Differentiate the orthogonality relation

$$\int p_{n+1}(x;t)p_n(x;t)e^{tx}\,d\mu(x) = 0$$

with respect to t to find

$$\int\left(\frac{d}{dt}p_{n+1}(x;t)\right)p_n(x;t)\,d\mu_t(x) + \int xp_{n+1}(x;t)p_n(x;t)\,d\mu_t(x) = 0,$$

which, by using (1.4), gives

$$\int\left(\frac{d}{dt}p_{n+1}(x;t)\right)p_n(x;t)\,d\mu_t(x) = -a_{n+1}.$$

Then we find

$$\frac{d}{dt}b_n = a_{n+1}^2 - a_n^2,$$

which is the required expression.

Exercise 9

The jump condition implies for the $(1, 1)$-entry of Y that

$$Y_{1,1}^+(x) = Y_{1,1}^-(x), \qquad x \in \mathbb{R},$$

which means that $Y_{1,1}$ has no jump and hence it is an entire function. The asymptotic condition for $Y_{1,1}$ gives $Y_{1,1}(z) = z^n + O(z^{n-1})$ as $z \to \infty$, which implies that $Y_{1,1}$ is a monic polynomial of degree n. Let us call this polynomial P_n, so that $Y_{1,1}(z) = P_n(z)$. At this point nothing more is known about this polynomial.

Next, consider the jump condition for the $(1, 2)$-entry of Y, then

$$Y_{1,2}^+(x) = Y_{1,1}^-(x)w(x) + Y_{1,2}^-(x), \qquad x \in \mathbb{R},$$

so that

$$Y_{1,2}^+(x) - Y_{1,2}^-(x) = P_n(x)w(x), \qquad x \in \mathbb{R}.$$

The asymptotic condition gives $Y_{1,2}(z) = O(z^{-n-1})$, hence the Sokhotsky–Plemelj formula gives

$$Y_{1,2}(z) = \frac{1}{2\pi i} \int_{-\infty}^{\infty} \frac{P_n(x)w(x)}{x - z} \, dx.$$

Now expand $1/(x - z)$ around $z = \infty$ as

$$\frac{1}{x - z} = -\sum_{k=0}^{m} \frac{x^k}{z^{k+1}} + O(z^{-m-2}),$$

then we find

$$Y_{1,2}(z) = -\sum_{k=0}^{m} \frac{1}{z^{k+1}} \frac{1}{2\pi i} \int_{-\infty}^{\infty} x^k P_n(x)w(x) \, dx + O(z^{-m-2}).$$

For $m = n - 1$ we then see that the asymptotic condition implies

$$\int_{-\infty}^{\infty} x^k P_n(x)w(x) \, dx = 0, \qquad 0 \le k \le n - 1,$$

which means that P_n is the (monic) orthogonal polynomial of degree n for the weight function w.

The reasoning for the second row of Y is similar. One finds that $Y_{2,1} = Q_{n-1}$

is a polynomial of degree $\leq n - 1$, but not necessarily a monic polynomial. For $Y_{2,2}$ the asymptotic condition gives

$$Y_{2,2}(z) = \frac{1}{2\pi i} \int_{-\infty}^{\infty} \frac{Q_{n-1}(x)w(x)}{x - z} \, dx = z^{-n} + O(z^{-n-1}), \qquad z \to \infty,$$

and one finds that

$$\int_{-\infty}^{\infty} x^k Q_{n-1}(x)w(x) \, dx = 0, \qquad 0 \leq k \leq n - 2,$$

and

$$\int_{-\infty}^{\infty} x^{n-1} Q_{n-1}(x)w(x) \, dx = -2\pi i.$$

This means that $Q_{n-1} = -2\pi i \gamma_{n-1}^2 P_{n-1}$, where P_{n-1} is the monic orthogonal polynomial of degree $n - 1$ for the weight function w.

Exercise 10

Let $y(t) = a_n(t^2)/t$, then

$$y'(t) = -\frac{a_n(t^2)}{t^2} + 2a_n'(t^2), \quad y''(t) = \frac{2a_n(t^2)}{t^3} - \frac{2a_n'(t^2)}{t} + 4t a_n''(t^2),$$

so that

$$a_n(t^2) = ty,$$

$$a_n'(t^2) = \frac{1}{2}\left(y' + \frac{y}{t}\right),$$

$$a_n''(t^2) = \frac{1}{4t}\left(y'' - \frac{y}{t^2} + \frac{y'}{t}\right).$$

Now use (4.23) for $s = t^2$ to find

$$a_n''(t^2) = \frac{(a_n'(t^2))^2}{a_n(t^2)} - \frac{a_n'(t^2)}{t^2} + (2n + \alpha + 1)\frac{a_n^2(t^2)}{t^4} + \frac{a_n^3(t^2)}{t^4} + \frac{\alpha}{t^2} - \frac{1}{a_n(t^2)},$$

which after the above substitutions becomes

$$\frac{1}{4t}\left(y'' - \frac{y}{t^2} + \frac{y'}{t}\right) = \frac{(y' + y/t)^2}{4ty} - \frac{y' + y/t}{2t^2} + \frac{(2n + \alpha + 1)y^2}{t^2} + \frac{y^3}{t} + \frac{\alpha}{t^2} - \frac{1}{ty}.$$

Multiply everything by $4t$ to find

$$y'' = \frac{(y')^2}{y} - \frac{y'}{t} + \frac{4(2n + \alpha + 1)y^2}{t} + 4y^3 + \frac{4\alpha}{t} - \frac{4}{y}.$$

This corresponds to (1.20) if the parameters in (1.20) are substituted for

$$\alpha \to 4(2n + \alpha + 1), \quad \beta \to 4\alpha, \quad \gamma \to 4, \quad \delta \to -4.$$

Exercise 11

One can use the results of Forrester and Witte for this special case $\mu = 0 = \omega$, but here is an alternative way, using a structure relation for the monic orthogonal polynomials, as was done by Magnus in [115]. Let $(\Phi_n)_{n \in \mathbb{N}}$ be the monic orthogonal polynomials on the unit circle for the weight (5.20). Define the polynomial Q by

$$Q(z) = (z - e^{i\pi\alpha})(z - e^{-i\pi\alpha}),$$

then we show that

$$Q(z)\Phi'_n(z) = a_n \Phi^*_{n-1}(z) + b_n \Phi_{n+1}(z) + c_n \Phi_n(z) + d_n \Phi_{n-1}(z), \qquad (A.17)$$

with a_n, b_n, c_n, d_n constants. Consider the integral

$$\frac{1}{2\pi} \int_{-\pi}^{\pi} Q(z)\Phi'_n(z) \overline{z^k} w(e^{i\theta}) \, d\theta$$

$$= \frac{B}{2\pi i} \int_{C_\alpha} Q(z)\Phi'_n(z) \frac{dz}{z^{k+1}} + \frac{A}{2\pi i} \int_{\mathbb{T} \setminus C_\alpha} Q(z)\Phi'_n(z) \frac{dz}{z^{k+1}},$$

where C_α is the arc $\{e^{i\theta}, -\alpha\pi < \theta < \alpha\pi\}$. Observe that Q vanishes at the endpoints of this arc. Integration by parts then gives

$$-\frac{B}{2\pi i} \int_{C_\alpha} \left(\frac{Q(z)}{z^{k+1}} \right)' \Phi_n(z) \, dz - \frac{A}{2\pi i} \int_{\mathbb{T} \setminus C_\alpha} \left(\frac{Q(z)}{z^{k+1}} \right)' \Phi_n(z) \, dz$$

$$= -\frac{1}{2\pi} \int_{-\pi}^{\pi} Q'(z)\Phi_n(z) \overline{z^k} w(e^{i\theta}) \, d\theta + \frac{k+1}{2\pi} \int_{-\pi}^{\pi} Q(z)\Phi_n(z) \overline{z^{k+1}} w(e^{i\theta}) \, d\theta.$$

$$(A.18)$$

Since Q is a polynomial of degree 2, the orthogonality relations for the polynomial Φ_n imply that the two integrals on the left vanish for $k = 1, 2, \ldots, n-2$, and hence the polynomial $Q\Phi'_n$ (which is of degree $n + 1$) is orthogonal to the polynomials $z, z^2, \cdots, z^{n-2}$. The four polynomials $\Phi_{n+1}, \Phi_n, \Phi_{n-1}$ and Φ^*_{n-1} are also orthogonal to these monomials and they are linearly independent, spanning the linear space of polynomials of degree $n + 1$ which are orthogonal to $z, z^2, \ldots, z^{n-2}$, hence (A.17) holds for certain constants a_n, b_n, c_n, d_n. The coefficients b_n and c_n can easily be found by comparing the coefficients of z^{n+1} and z^n: if $\Phi_n(z) = z^n + \beta_n z^{n-1} + \cdots$, then

$$b_n = n, \quad c_n = -2n \cos \pi\alpha - \beta_n - n r_{n+1} r_n.$$

For a_n we integrate (A.17) with the weight w to find

$$\frac{1}{2\pi} \int_{-\pi}^{\pi} Q(z)\Phi'_n(z) w(z) \, d\theta = \frac{a_n}{2\pi} \int_{-\pi}^{\pi} \Phi^*_n(z) w(z) \, d\theta.$$

The left-hand side is the integral in (A.18) for $k = 0$, hence

$$\frac{1}{2\pi} \int_{-\pi}^{\pi} Q(z)\Phi_n'(z)w(e^{i\theta})\, d\theta = -\frac{1}{2\pi} \int_{-\pi}^{\pi} Q'(z)\Phi_n(z)w(e^{i\theta})\, d\theta$$

$$+ \frac{k+1}{2\pi} \int_{-\pi}^{\pi} Q(z)\Phi_n(z)\bar{z}w(e^{i\theta})\, d\theta$$

$$= -\frac{1}{2\pi} \int_{-\pi}^{\pi} z\Phi_n(z)w(e^{i\theta})\, d\theta = \frac{r_{n+1}}{\kappa_n^2},$$

where we use the recurrence $\Phi_{n+1}(z) = z\Phi_n(z) + r_{n+1}\Phi_n^*(z)$ for the last equality. On the other hand,

$$\frac{a_n}{2\pi} \int_{-\pi}^{\pi} \Phi_n^*(z)w(z)\, d\theta = \frac{a_n}{\kappa_{n-1}^2},$$

and hence it follows that

$$a_n = r_{n+1}\frac{\kappa_{n-1}^2}{\kappa_n^2} = r_{n+1}(1 - r_n^2),$$

where we used (3.3). For d_n we multiply (A.17) by $\overline{z^{n-1}}$ and integrate to find

$$\frac{1}{2\pi} \int_{-\pi}^{\pi} Q(z)\Phi_n'(z)\overline{z^{n-1}}w(e^{i\theta})\, d\theta = \frac{d_n}{\kappa_{n-1}^2}.$$

The left-hand side is the integral in (A.18) for $k = n - 1$, and hence the orthogonality gives

$$\frac{1}{2\pi} \int_{-\pi}^{\pi} Q(z)\Phi_n'(z)\overline{z^{n-1}}w(e^{i\theta})\, d\theta = -\frac{1}{2\pi} \int_{-\pi}^{\pi} Q'(z)\Phi_n(z)\overline{z^{n-1}}w(e^{i\theta})\, d\theta$$

$$+ \frac{n}{2\pi} \int_{-\pi}^{\pi} Q(z)\Phi_n(z)\overline{z^n}w(e^{i\theta})\, d\theta$$

$$= \frac{n}{2\pi} \int_{-\pi}^{\pi} \Phi_n(z)\overline{z^n}w(e^{i\theta})\, d\theta = \frac{n}{\kappa_n^2},$$

so that $d_n = n(1 - r_n^2)$. Hence (A.17) becomes

$$Q(z)\Phi_n'(z) = r_{n+1}(1 - r_n^2)\Phi_{n-1}^*(z) + n\Phi_{n+1}(z)$$

$$- (2n\cos\pi\alpha + \beta_n + nr_{n+1}r_n)\Phi_n(z) + n(1 - r_n^2)\Phi_{n-1}(z).$$

If we evaluate this at $z = 0$ and use $\Phi_n'(0) = r_{n-1} + r_n\beta_{n-1} = r_{n-1}(1 - r_n^2) + r_n\beta_n$, then the required recurrence relation (5.21) follows. The initial conditions $r_0 = 1$ and $\beta_0 = 0$ are immediate, and $r_1 = -m_1/m_0$, where

$$m_0 = \frac{1}{2\pi} \int_{-\pi}^{\pi} w(e^{i\theta})\, d\theta = \alpha B + (1 - \alpha)A,$$

$$m_1 = \frac{1}{2\pi} \int_{-\pi}^{\pi} e^{i\theta}w(e^{i\theta})\, d\theta = (B - A)\frac{\sin\pi\alpha}{\pi}.$$

Exercise 12

Let us denote the generating function for the polynomials $(p_n)_{n \in \mathbb{N}}$ by $G(z, \lambda)$ so that

$$\sum_{k=0}^{\infty} p_k(z)\lambda^k = G(z, \lambda) = \exp(z\lambda - \frac{4}{3}\lambda^3).$$

Differentiation with respect to z gives

$$\sum_{k=0}^{\infty} p_k'(z)\lambda^k = \frac{\partial}{\partial z}G(z, \lambda) = \lambda G(z, \lambda) = \sum_{j=0}^{\infty} p_j(z)\lambda^{j+1}.$$

Changing j to $k - 1$ in this last sum gives

$$\sum_{k=0}^{\infty} p_k'(z)\lambda^k = \sum_{k=1}^{\infty} p_{k-1}(z)\lambda^k,$$

and by identifying terms of λ^n one finds that $p_n'(z) = p_{n-1}(z)$. This immediately also gives $p_n^{(k)}(z) = p_{n-k}(z)$ by induction. Next, by differentiating the generating function with respect to λ one finds

$$\sum_{k=1}^{\infty} k p_k(z)\lambda^{k-1} = \frac{\partial}{\partial \lambda}G(z, \lambda) = (z - 4\lambda^2)G(z, \lambda). \tag{A.19}$$

The latter expression is also

$$(z - 4\lambda^2)G(z, \lambda) = \sum_{n=0}^{\infty} z p_n(z)\lambda^n - 4\sum_{j=0}^{\infty} p_j(z)\lambda^{j+2},$$

and by changing j to $n - 2$ in the last sum, and k to $n + 1$ in the first sum in (A.19), we find

$$\sum_{n=0}^{\infty}(n + 1)p_{n+1}(z)\lambda^n = \sum_{n=0}^{\infty} z p_n(z)\lambda^n - 4\sum_{n=2}^{\infty} p_{n-2}(z)\lambda^n.$$

By identifying the terms of λ^n we then find

$$(n + 1)p_{n+1}(z) = z p_n(z) - 4p_{n-2}(z),$$

which is the required recurrence relation. The differential equation simply follows by changing n to $n - 1$ and using $p_n'(z) = p_{n-1}(z)$ and $p_n'''(z) = p_{n-3}(z)$.

Exercise 13

This is Lemma 4.4 in [127], which is proved in Appendix B of that paper. The identities (6.21) follow from the identity of Laguerre polynomials [123, Eq. 18.9.13]

$$L_k^{(\alpha+1)}(x) - L_{k-1}^{(\alpha+1)}(x) = L_k^{(\alpha)}(x),$$

which can be proved by using the generating function (6.17). One has

$$\sum_{k=0}^{\infty} L_k^{(\alpha)}(x) z^k = (1-z)^{-\alpha-1} \exp\left(-x\frac{z}{1-z}\right)$$

$$= (1-z)(1-z)^{-\alpha-2} \exp\left(-x\frac{z}{1-z}\right)$$

$$= \sum_{k=0}^{\infty} L_k^{(\alpha+1)}(x) z^k - \sum_{k=0}^{\infty} L_k^{(\alpha+1)}(x) z^{k+1}$$

and comparing powers of z gives the required identity.

For the various subdeterminants (minors) of D, we see that the results for $D_{[m]}^{[1]}$, $D_{[m+1]}^{[1]}$ and $D_{[m,m+1]}^{[1,m+n+1]}$ follow immediately from (6.20) by deleting the relevant rows and columns. If we replace in (6.20) column j by column j minus column $j-1$ (for $j = 2, \ldots, n+m$) and use (6.21), then we find

$$R_{m,n}(\mu+1) = (-1)^m \begin{vmatrix} -q_1^{(r+1)} & q_1^{(r)} & q_0^{(r)} & \cdots & q_{-m-n+4}^{(r)} & q_{-m-n+3}^{(r)} \\ -q_3^{(r+1)} & q_3^{(r)} & q_2^{(r)} & \cdots & q_{-m-n+4}^{(r)} & q_{-m-n+5}^{(r)} \\ \vdots & \vdots & \vdots & \cdots & \vdots & \vdots \\ -q_{2m-1}^{(r+1)} & q_{2m-1}^{(r)} & q_{2m-2}^{(r)} & \cdots & q_{m-n}^{(r)} & q_{m-n+1}^{(r)} \\ p_{n-m}^{(r+1)} & p_{n-m+1}^{(r)} & p_{n-m+2}^{(r)} & \cdots & p_{2n-2}^{(r)} & p_{2n-1}^{(r)} \\ p_{n-m-2}^{(r+1)} & p_{n-m-1}^{(r)} & p_{n-m}^{(r)} & \cdots & p_{2n-4}^{(r)} & p_{2n-3}^{(r)} \\ \vdots & \vdots & \vdots & \cdots & \vdots & \vdots \\ p_{-n-m+2}^{(r+1)} & p_{-n-m+3}^{(r)} & p_{-n-m+4}^{(r)} & \cdots & p_0^{(r)} & p_1^{(r)} \end{vmatrix},$$

from which we see that $D = (-1)^m R_{m,n+1}(\mu+1)$. We can use $p_0^{(r)} = 1$ and

$p_k^{(r)} = 0$ for $k < 0$ to add a row and a column at the end in (6.20) to find

$$
R_{m,n}(\mu) =
\begin{vmatrix}
q_1^{(r)} & q_0^{(r)} & \cdots & q_{-m-n+3}^{(r)} & q_{-m-n+2}^{(r)} & q_{-m-n+1}^{(r)} \\
q_3^{(r)} & q_2^{(r)} & \cdots & q_{-m-n+5}^{(r)} & q_{-m-n+4}^{(r)} & q_{-m-n+3}^{(r)} \\
\vdots & \vdots & \cdots & \vdots & \vdots & \vdots \\
q_{2m-1}^{(r)} & q_{2m-2}^{(r)} & \cdots & q_{m-n+1}^{(r)} & q_{m-n}^{(r)} & q_{m-n-1}^{(r)} \\
p_{n-m}^{(r)} & p_{n-m+1}^{(r)} & \cdots & p_{2n-2}^{(r)} & p_{2n-1}^{(r)} & p_{2n}^{(r)} \\
p_{n-m-2}^{(r)} & p_{n-m-1}^{(r)} & \cdots & p_{2n-4}^{(r)} & p_{2n-3}^{(r)} & p_{2n-2}^{(r)} \\
\vdots & \vdots & \cdots & \vdots & \vdots & \vdots \\
p_{-n-m+2}^{(r)} & p_{-n-m+3}^{(r)} & \cdots & p_0^{(r)} & p_1^{(r)} & p_2^{(r)} \\
p_{-n-m}^{(r)} & p_{-n-m+1}^{(r)} & \cdots & p_{-2}^{(r)} & p_{-1}^{(r)} & p_0^{(r)}
\end{vmatrix}.
$$

Again replacing column j by column j minus column $j-1$ (for $j = 2, \ldots, n+m$) and using (6.21) gives

$$
R_{m,n}(\mu + 1) = (-1)^m
\begin{vmatrix}
-q_1^{(r+1)} & q_1^{(r)} & \cdots & q_{-m-n+4}^{(r)} & q_{-m-n+3}^{(r)} & q_{-m-n+2}^{(r)} \\
-q_3^{(r+1)} & q_3^{(r)} & \cdots & q_{-m-n+6}^{(r)} & q_{-m-n+5}^{(r)} & q_{-m-n+4}^{(r)} \\
\vdots & \vdots & \cdots & \vdots & \vdots & \vdots \\
-q_{2m-1}^{(r+1)} & q_{2m-1}^{(r)} & \cdots & q_{m-n+2}^{(r)} & q_{m-n+1}^{(r)} & q_{m-n}^{(r)} \\
p_{n-m}^{(r+1)} & p_{n-m+1}^{(r)} & \cdots & p_{2n-2}^{(r)} & p_{2n-1}^{(r)} & p_{2n}^{(r)} \\
p_{n-m-2}^{(r+1)} & p_{n-m-1}^{(r)} & \cdots & p_{2n-4}^{(r)} & p_{2n-3}^{(r)} & p_{2n-2}^{(r)} \\
\vdots & \vdots & \cdots & \vdots & \vdots & \vdots \\
p_{-n-m+2}^{(r+1)} & p_{-n-m+3}^{(r)} & \cdots & p_0^{(r)} & p_1^{(r)} & p_2^{(r)} \\
p_{-n-m}^{(r+1)} & p_{-n-m+1}^{(r)} & \cdots & p_{-2}^{(r)} & p_{-1}^{(r)} & p_0^{(r)}
\end{vmatrix},
$$

from which we find the formulas for $D_{[m]}^{[m+n+1]}$ and $D_{[m+1]}^{[m+n+1]}$.

Exercise 14

This is an exercise in manipulating determinants. Two properties are needed. First, if A is a matrix with entries depending on x, then

$$
\frac{d}{dx} \det A = \sum_{k=1}^{n} \det A_k = \sum_{k=1}^{n} \det A^k,
$$

where A_k is the matrix A but with the elements in the k-th row replaced by their derivatives, and A^k with the k-th column replaced by its derivatives. The second property is the *Desnanot–Jacobi identity* or the Lewis Carroll[1] formula. Denote by $A^{[k,\ell]}_{[i,j]}$ (with $i < j$ and $k < \ell$) the matrix A with the rows i and j and the columns k, ℓ removed, and similar for $A^{[k]}_{[i]}$ with row i and column k removed, then

$$\det A \det A^{[k,\ell]}_{[i,j]} = \det A^{[k]}_{[i]} \det A^{[\ell]}_{[j]} - \det A^{[\ell]}_{[i]} \det A^{[k]}_{[j]}.$$

Take for A the matrix in the definition for τ_{n+1}

$$A = \begin{pmatrix} \varphi & \varphi' & \cdots & \varphi^{(n-1)} & \varphi^{(n)} \\ \varphi' & \varphi'' & \cdots & \varphi^{(n)} & \varphi^{(n+1)} \\ \vdots & \vdots & \cdots & \vdots & \vdots \\ \varphi^{(n-1)} & \varphi^{(n)} & \cdots & \varphi^{(2n-2)} & \varphi^{(2n-1)} \\ \varphi^{(n)} & \varphi^{(n+1)} & \cdots & \varphi^{(2n-1)} & \varphi^{(2n)} \end{pmatrix}.$$

If we differentiate $\det A$, then the derivative of each row corresponds to the next row, and hence each A_k with $1 \le k \le n$ has two equal rows, so that $\det A_k$ vanishes. Hence only A_{n+1} has a contribution and $(\det A)' = \det A_{n+1}$. In a similar way one also has $(\det A)' = \det A^{n+1}$. In terms of the functions τ_n this gives

$$\det A = \tau_{n+1}, \quad \det A^{[n+1]}_{[n+1]} = \tau_n, \quad \det A^{[n,n+1]}_{[n,n+1]} = \tau_{n-1},$$

and for the derivatives

$$\det A^{[n+1]}_{[n]} = \tau'_n = \det A^{[n]}_{[n+1]}, \quad \det A^{[n]}_{[n]} = \tau''_n.$$

Note that this is true for any function φ which has sufficiently many derivatives. The Desnanot–Jacobi identity then gives

$$\tau_{n+1}\tau_{n-1} = \tau_n \tau''_n - (\tau'_n)^2.$$

One easily verifies the initial conditions $\tau_0 = 1$ and $\tau_1 = \varphi$.

Exercise 15

By taking the derivative in the Riccati equation (6.29) one finds

$$y'' = 2yy' + 2y + 2xy'.$$

---

[1] Lewis Carroll is well known as the author of *Alice's Adventures in Wonderland* and *Through the Looking Glass*, but his real name was Charles Lutwidge Dodgson and he was a mathematician and logician.

The Painlevé IV equation is

$$y'' = \frac{(y')^2}{2y} + \frac{3}{2}y^3 + 4xy^2 + 2(x^2 - \alpha) + \frac{\beta}{y},$$

hence one needs to show that

$$2yy' + 2y + 2xy' = \frac{(y')^2}{2y} + \frac{3}{2}y^3 + 4xy^2 + 2(x^2 - \alpha) + \frac{\beta}{y}.$$

Replace y' everywhere by $y^2 + 2xy - 2(1+\alpha)$, which follows from (6.29), then almost all the terms cancel and the only terms left are $\beta = -2(1+\alpha)^2$.

A similar reasoning holds for (6.30), which after derivation gives

$$y'' = -2yy' - 2y - 2xy'.$$

Comparing with Painlevé IV and replacing all y' by $-y^2 - 2xy - 2(1-\alpha)$ gives, after many cancellations, that $\beta = -2(1-\alpha)^2$.

Exercise 16

We will show that a solution of Painlevé II

$$y''(x) = 2y^3(x) + xy(x) - 2\alpha - \frac{1}{2}$$

in the transformation formulas $u(s) = 2^{-1/3}U(x)$ and $U(x) = y^2(x) + y'(x) + x/2$, with $x = -2^{1/3}s$, will give a solution of Painlevé XXXIV given by (7.16). The computations for the inverse transformation are similar and we encourage the reader to make them anyway. We have

$$u'(s) = -U'(x) = -2y(x)y'(x) - y''(x) - \frac{1}{2}$$

and if we use the differential equation for y, then this gives

$$u'(s) = -2y(x)y'(x) - 2y^3(x) - xy(x) + 2\alpha. \tag{A.20}$$

Differentiate again to find

$$u''(s) = 2^{1/3}U''(x) = 2[y'(x)]^2 + 2y(x)y'(x) + 6y'(x)y^2(x) + y(x) + xy'(x),$$

and by replacing y'' one has

$$u''(s) = 2^{1/3}(2[y'(x)]^2 + 2y(x)[2y^3(x) + xy(x) - 2\alpha] + 6y'(x)y^2(x) + xy'(x). \tag{A.21}$$

Observe that

$$[U'(x)]^2 - (2\alpha)^2 = (U'(x) + 2\alpha)(U'(x) - 2\alpha)$$
$$= 2y(x)U(x)[2y(x)y'(x) + 2y^3(x) + xy(x) - 4\alpha],$$

so that

$$4u^2(s) + 2su(s) + \frac{[u'(s)]^2 - (2\alpha)^2}{2u(s)} = 2^{4/3}[y^2(x) + y'(x) - 2^{-2/3}s]^2$$
$$+ 2^{2/3}s[y^2(x) + y'(x) - 2^{-2/3}s] + 2^{1/3}y(x)[2y(x)y'(x) + 2y^3(x) - 2^{1/3}sy(x) - 4\alpha].$$

Simple comparison with (A.21) shows that u indeed satisfies (7.16).

References

[1] H. Airault, *Rational solutions of Painlevé equations*, Stud. Appl. Math. **61** (1979), no. 1, 31–53.

[2] S.M. Alsulami, P. Nevai, J. Szabados, W. Van Assche, *A family of nonlinear difference equations: existence, uniqueness, and asymptotic behavior of positive solutions*, J. Approx. Theory **193** (2015), 39–55.

[3] R. Askey, J. Wilson, *Some basic hypergeometric orthogonal polynomials that generalize Jacobi polynomials*, Memoirs of the American Mathematical Society **54**, no. 319 (1985), Amer. Math. Soc., Providence, RI.

[4] M.R. Atkin, T. Claeys, F. Mezzadri, *Random matrix ensembles with singularities and a hierarchy of Painlevé III equations*, Int. Math. Res. Not. **2016** (2016), no. 8, 2320–2375.

[5] J. Baik, P. Deift, K. Johansson, *On the distribution of the length of the longest increasing subsequence of random permutations*, J. Amer. Math. Soc. **12** (1999), no. 4, 1119–1178.

[6] E. Basor, Y. Chen, T. Ehrhardt, *Painlevé V and time dependent Jacobi polynomials*, J. Phys. A: Math. Theor. **43** (2010), no. 1, 015204.

[7] M. Bertola, T. Bothner, *Zeros of large degree Vorob'ev-Yablonski polynomials via a Hankel determinant identity*, Int. Math. Res. Not. 2015, no. 19, 9330–9399.

[8] Ph. Biane, *Orthogonal polynomials on the unit circle, q-gamma weights, and discrete Painlevé equations*, Mosc. Math. J. **14** (2014), no. 1, 1–27.

[9] P.M. Bleher, A. Deaño, *Topological expansion in the cubic random matrix model*, Int. Math. Res. Not. (2013), no. 12, 2699–2755.

[10] P. Bleher, A. Its, *Semiclassical asymptotics of orthogonal polynomials, Riemann-Hilbert problems, and universality in the random matrix model*, Ann. of Math. (2) **150** (1999), no. 1, 185–266.

[11] P. Bleher, A. Its, *Double scaling limit in the random matrix model: the Riemann-Hilbert approach*, Commun. Pure Appl. Math. **56** (2003), no. 4, 433–516.

[12] L. Boelen, G. Filipuk, W. Van Assche, *Recurrence coefficients of generalized Meixner polynomials and Painlevé equations*, J. Phys. A: Math. Theor. **44**, number 3 (2011), 035202 (19 pp.).

[13] L. Boelen, C. Smet, W. Van Assche, *q-Discrete Painlevé equations for recurrence coefficients of modified q-Freud orthogonal polynomials*, J. Difference Equations Appl. **16** (2010), no. 1, 37–53.

[14] L. Boelen, W. Van Assche, *Discrete Painlevé equations for recurrence coefficients of semiclassical Laguerre polynomials*, Proc. Amer. Math. Soc. **138**, no. 4 (2010), 1317–1331.

[15] L. Boelen, W. Van Assche, *Variations of Stieltjes-Wigert and q-Laguerre polynomials and their recurrence coefficients*, J. Approx. Theory **193** (2015), 56–73.

[16] A. Bogatskiy, T. Claeys, A. Its, *Hankel determinant and orthogonal polynomials for a Gaussian weight with a discontinuity at the edge*, Comm. Math. Phys. **347** (2016), no. 1, 127–162.

[17] S. Bonan, P. Nevai, *Orthogonal polynomials and their derivatives, I*, J. Approx. Theory, **40** (1984), no. 2, 134–147.

[18] N. Bonneux, A.B.J. Kuijlaars, Exceptional Laguerre polynomials, arXiv:1708.03106 [math.CA] (August 2017).

[19] T. Bothner, P.D. Miller, Y. Sheng, *Large degree asymptotics of rational solutions of the Painlevé-III equation* (in preparation).

[20] L. Brightmore, F. Mezzadri, M.Y. Mo, *A matrix model with singular weight and Painlevé III*, Comm. Math. Phys. **333** (2015), 1317–1364.

[21] R.J. Buckingham, *Large-degree asymptotics of rational Painlevé-IV functions associated to generalized Hermite polynomials*, arXiv:1706.09005 [math-ph] (June 2017).

[22] R.J. Buckingham, P.D. Miller, *Large-degree asymptotics of rational Painlevé-II functions: noncritical behaviour*, Nonlinearity **27** (2014), no. 10, 2489–2578.

[23] R.J. Buckingham, P.D. Miller, *Large-degree asymptotics of rational Painlevé-II functions: critical behaviour*, Nonlinearity **28** (2015), no. 6, 1539–1596.

[24] Y. Chen, D. Dai, *Painlevé V and a Pollaczek-Jacobi type orthogonal polynomials*, J. Approx. Theory **162** (2010), no. 12, 2149–2167.

[25] Y. Chen, M.E.H. Ismail, *Ladder operators and differential equations for orthogonal polynomials*, J. Phys. A: Math. Gen. **30** (1997), no. 22, 7817–7829.

[26] Y. Chen, M.E.H. Ismail, *Ladder operators for q-orthogonal polynomials*, J. Math. Anal. Appl. **345** (2008), no. 1, 1–10.

[27] Y. Chen, A. Its, *Painlevé III and a singular linear statistics in Hermitian random matrix ensembles, I*, J. Approx. Theory **162** (2010), no. 2, 270–297.

[28] Y. Chen, L. Zhang, *Painlevé VI and the unitary Jacobi ensembles*, Stud. Math. **125** (2010), no. 1, 91–112.

[29] T. Claeys, B. Fahs, *Random matrices with merging singularities and the Painlevé V equation*, Symmetry Integr. Geom. Meth. Appl. (SIGMA) **12** (2016), 031, 44 pp.

[30] T. Claeys, A. Its, I. Krasovsky, *Emergence of a singularity for Toeplitz determinants and Painlevé V*, Duke Math. J. **160** (2011), no. 2, 207–262.

[31] T. Claeys, I. Krasovsky, *Toeplitz determinants with merging singularities*, Duke Math. J. **164** (2015), no. 15, 2897–2987.

[32] T. Claeys, A.B.J. Kuijlaars, *Universality of the double scaling limit in random matrix models*, Commun. Pure Appl. Math. **59** (2006), no. 11, 1573–1603.

[33] T. Claeys, A.B.J. Kuijlaars, *Universality in unitary random matrix ensembles when the soft end meets the hard edge*, in "Integrable Systems and Random Matrices", Contemp. Math. **458**, Amer. Math. Soc., Providence, RI, 2008, pp. 265–279.

[34] T. Claeys, A.B.J. Kuijlaars, M. Vanlessen, *Multi-critical unitary random matrix ensembles and the general Painlevé II equation*, Ann. of Math. (2) **168** (2008), no. 2, 601–641.

[35] T. Claeys, M. Vanlessen, *Universality of a double scaling limit near singular edge points in random matrix models*, Commun. Math. Phys. **273** (2007), no. 2, 499–532.

[36] P.A. Clarkson, *The third Painlevé equation and associated special polynomials*, J. Phys. A: Math. Gen. **36** (2003), no. 36, 9507–9532.

[37] P.A. Clarkson, *The fourth Painlevé equation and associated special polynomials*, J. Math. Phys. **44** (2003), no. 11, 5350–5374.

[38] P.A. Clarkson, *Special polynomials associated with rational solutions of the fifth Painlevé equation*, J. Comput. Appl. Math. **178** (2005), no. 1–2, 111–129.

[39] P.A. Clarkson, *Painlevé equations — nonlinear special functions*, in "Orthogonal Polynomials and Special Functions" (F. Marcellán, W. Van Assche, eds.), Lecture Notes in Mathematics **1883**, Springer, Berlin, 2006. pp. 331–411.

[40] P.A. Clarkson, *Special polynomials associated with rational and algebraic solutions of the Painlevé equations*, in "Théories asymptotiques et équations de Painlevé", Sémin. Cong. **14**, Soc. Math. France, Paris, 2006, pp. 21–52.

[41] P.A. Clarkson, *Recurrence coefficients for discrete orthogonal polynomials and the Painlevé equations*, J. Phys. A: Math. Theor. **46** (2013), no. 18, 185205 (18 pp.).

[42] P.A. Clarkson, K. Jordaan, *The relationship between semiclassical Laguerre polynomials and the fourth Painlevé equation*, Constr. Approx. **39** (2014), no. 1, 223–254.

[43] P.A. Clarkson, K. Jordaan, A. Kelil, *A generalized Freud weight*, Stud. Appl. Math. **136** (2016), no. 3, 288–320.

[44] P.A. Clarkson, A.F. Loureiro, W. Van Assche, *Unique positive solution for an alternative discrete Painlevé I equation*, J. Difference Equations Appl. **22** (2016), no. 5, 656–675.

[45] P.A. Clarkson, E.L. Mansfield, *The second Painlevé equation, its hierarchy and associated special polynomials*, Nonlinearity **16** (2003), no. 3, R1–R26.

[46] R. Conte, M. Musette, *The Painlevé handbook*, Springer, Dordrecht, 2008.

[47] D. Dai, *Asymptotics of orthogonal polynomials and the Painlevé transcendents*, arXiv:1608.04513 [math.CA].

[48] D. Dai, A.B.J. Kuijlaars, *Painlevé IV asymptotics for orthogonal polynomials with respect to a modified Laguerre weight*, Stud. Appl. Math. **122** (2009), no. 1, 29–83.

[49] D. Dai, L. Zhang, *Painlevé VI and Hankel determinants for the generalized Jacobi weight*, J. Phys. A: Math. Theor. **43** (2010), 055207, 14 pp.

[50] A. Deaño, D. Huybrechs, A.B.J. Kuijlaars, *Asymptotic zero distribution of complex orthogonal polynomials associated with Gaussian quadrature*, J. Approx. Theory **162** (2010), 2202–2224.

[51] P.A. Deift, *Orthogonal polynomials and random matrices: a Riemann-Hilbert approach*, Courant Lecture Notes in Mathematics, vol. 3, Courant Institute of Mathematical Sciences, New York; American Mathematical Society, Providence, RI, 1999.

[52] P. Deift, A. Its, I. Krasovsky, *Asymptotics of Toeplitz, Jankel, and Toeplitz+Hankel determinants with Fisher-Hartwig singularities*, Ann. of Math. (2) **174** (2011), 1243–1299.

[53] P. Deift, T. Kriecherbauer, K.T.-R. McLaughlin, *New results on the equilibrium measure for logarithmic potentials in the presence of an external field*, J. Approx. Theory **95** (1998), no. 3, 388–475.

[54] P. Deift, X. Zhou, *A steepest descent method for oscillatory Riemann-Hilbert problems. Asymptotics for the MKdV equation*, Ann. of Math. (2) **137** (1993), 295–368.

[55] P.A. Deift, X. Zhou, *Asymptotics for the Painlevé II equation*, Comm. Pure Appl. Math. **48** (1995), no. 3, 277–337.

[56] D.K. Dimitrov, Y.C. Lun, *Monotonicity, interlacing and electrostatic interpretation of zeros of exceptional Jacobi polynomials*, J. Approx. Theory **181** (2014), 18–29.

[57] M. Duits, A.B.J. Kuijlaars, *Painlevé I asymptotics for orthogonal polynomials with respect to a varying quartic weight*, Nonlinearity **19** (2006), no. 10, 2211–2245.

[58] A.J. Durán, *Exceptional Charlier and Hermite orthogonal polynomials*, J. Approx. Theory **182** (2014), 29–58.

[59] A.J. Durán, *Exceptional Meixner and Laguerre orthogonal polynomials*, J. Approx. Theory **184** (2014), 176–208.

[60] A.J. Durán, *Exceptional Hahn and Jacobi polynomials*, J. Approx. Theory **214** (2017), 9–48.

[61] A.J. Durán, M. Pérez, *Admissibility condition for exceptional Laguerre polynomials*, J. Math. Anal. Appl. **424** (2015), no. 2, 1042–1053.

[62] T. Ehrhardt, *A status report on the asymptotic behavior of Toeplitz determinants with Fisher-Hartwig singularities*, in "Recent Advances in Operator Theory (Groningen, 1998)", Oper. Theory Adv. Appl. **124**, Birkhäuser, Basel, 2001, pp. 217–241.

[63] G. Felder, A.D. Hemery, A.P. Veselov, *Zeros of Wronskians of Hermite polynomials and Young diagrams*, Physica D **241** (2012), 2131–2137.

[64] G. Filipuk, W. Van Assche, *Recurrence coefficients of a new generalization of the Meixner polynomials*, Symmetry Integr. Geom. Meth. Appl. (SIGMA) **7** (2011), 068.

[65] G. Filipuk, W. Van Assche, L. Zhang, *The recurrence coefficients of semi-classical Laguerre polynomials and the fourth Painlevé equation*, J. Phys. A: Math. Theor. **45**, number 20 (2012), 205201, 13 pp.

[66] G. Filipuk, W. Van Assche, L. Zhang, *Multiple orthogonal polynomials associated with an exponential cubic weight*, J. Approx. Theory **190** (2015), 1–25.

[67] M.E. Fisher, R.E. Hartwig, *Toeplitz determinants: some applications, theorems, and conjectures*, Adv. Chem. Phys. **15** (1968), 333–353.

[68] A.S. Fokas, A.R. Its, A.A. Kapaev, V.Yu. Novokshenov, *Painlevé transcendents: the Riemann-Hilbert approach*, AMS Mathematical Surveys and Monographs, vol. 128, Amer. Math. Soc., Providence, RI, 2006.

[69] A.S. Fokas, A.R. Its, A.V. Kitaev, *Discrete Painlevé equations and their appearance in quantum gravity*, Commun. Math. Phys. **142** (1991), no. 2, 313–344.

[70] A.S. Fokas, A.R. Its, A.V. Kitaev, *The isomonodromy approach to matrix models in 2D quantum gravity*, Comm. Math. Phys. **147** (1992), no. 2, 395–430.

[71] P.J. Forrester, N.S. Witte, *Application of the τ-function theory of Painlevé equations to random matrices: P$_V$, P$_{III}$, the LUE, JUE, and CUE*, Comm. Pure Appl. Math. **55** (2002), no. 6, 679–727.

[72] P.J. Forrester, N.S. Witte, *Discrete Painlevé equations, orthogonal polynomials on the unit circle, and N-recurrences for averages over U(N) — P$_{III'}$ and P$_V$ τ-functions*, Int. Math. Res. Not. **2004** (2004), no. 4, 160–183.

[73] P.J. Forrester, N.S. Witte, *Application of the τ-function theory of Painlevé equations to random matrices: P$_{VI}$, the JUE, CyUE, cJUE and scaled limits*, Nagoya Math. J. **174** (2004), 29–114.

[74] P.J. Forrester, N.S. Witte, *Discrete Painlevé equations for a class of PVI τ-functions given as U(N) averages*, Nonlinearity **18** (2005), 2061–2088.

[75] P.J. Forrester, N.S. Witte, *Bi-orthogonal polynomials on the unit circle, regular semi-classical weights and integrable systems*, Constr. Approx. **24** (2006), 201–237.

[76] M. Foupouagnigni, W. Van Assche, *Analysis of non-linear recurrence relations for the recurrence coefficients of generalized Charlier polynomials*, J. Nonlinear Math. Phys. **10** Supplement 2 (2003), 231–237.

[77] G. Freud, *On the coefficients in the recursion formulae of orthogonal polynomials*, Proc. Roy. Irish Acad. Sect. A **76** (1976), no. 1, 1–6.

[78] D. Gómez-Ullate, Y. Grandati, R. Milson, *Rational extensions of the quantum harmonic oscillator and exceptional Hermite polynomials*, J. Phys. A: Math. Gen. **47** (2014), no. 1, 015203, 27 pp.

[79] D. Gómez-Ullate, F. Marcellán, R. Milson, *Asymptotic and interlacing properties of zeros of exceptional Jacobi and Laguerre polynomials*, J. Math. Anal. Appl. **399** (2013), no. 2, 480–495.

[80] B. Grammaticos, A. Ramani, *Discrete Painlevé equations: a review*, Lecture Notes in Physics **644** (2004), 245–321.

[81] Y. Grandati, C. Quesne, *Disconjugacy, regularity of multi-index rationally extended potentials, and Laguerre exceptional polynomials*, J. Math. Phys. **54** (2013), no. 7, 073512, 13 pp.

[82] V.I. Gromak, I. Laine, S. Shimomura, *Painlevé differential equations in the complex plane*, de Gruyter Studies in Mathematics, vol. 28, Walter de Gruyter, Berlin, 2002.

[83] V.I. Gromak, N.A. Lukashevich, *Special classes of solutions of Painlevé's equations*, Diff. Eq. **18** (1982), 317–326.

[84] S.P. Hastings, J.B. McLeod, *A boundary value problem associated with the second Painlevé transcendent and the Korteweg-de Vries equation*, Arch. Rational Mech. Anal. **73** (1980) 31–51.

[85] J. Hietarinta, N. Joshi, F.W. Nijhoff, *Discrete systems and integrability*, Cambridge Texts in Applied Mathematics, Cambridge University Press, 2016.

[86] M. Hisakado, *Unitary matrix models and Painlevé III*, Mod. Phys. Lett. **A11** (1996), 3001–3010.

[87] M.N. Hounkonnou, C. Hounga, A. Ronveaux, *Discrete semi-classical orthogonal polynomials: generalized Charlier*, J. Comput. Appl. Math. **114** (2000), 361–366.

[88] E.L. Ince, *Ordinary differential equations*, Longmans, Green and Co., London, 1927; Dover Publications, New York, 1956.

[89] M.E.H. Ismail, *The Askey–Wilson operator and summation theorems*, in "Mathematical Analysis, Wavelets, and Signal Processing" (Cairo, 1994), Contemp. Math. **190**, Amer. Math. Soc., Providence, RI, 1995, pp. 171–178.

[90] M.E.H. Ismail, *Difference equations and quantized discriminants for q-orthogonal polynomials*, Adv. Appl. Math. **30** (2003), 562–589.

[91] M.E.H. Ismail, *Classical and quantum orthogonal polynomials in one variable*, Encyclopedia of Mathematics and its Applications **98**, Cambridge University Press, 2005.

[92] M.E.H. Ismail, I. Nikolova, P. Simeonov, *Difference equations and discriminants for discrete orthogonal polynomials*, Ramanujan J. **8** (2004), 475–502.

[93] A.R. Its, A.B.J. Kuijlaars, J. Östensson, *Critical edge behavior in unitary random matrix ensembles and the thirty-fourth Painlevé transcendent*, Int. Math. Res. Not. (2008), no. 9, rnn017, 67 pp.

[94] N. Joshi, A.V. Kitaev, *On Boutroux's tritronquée solutions of the first Painlevé equation*, Studies in Applied Mathematics **107** (2001), 253–291.

[95] K. Kajiwara, T. Masuda, *On the Umemura polynomials for the Painlevé III equation*, Phys. Lett. A **260** (1999), no. 6, 462–467.

[96] K. Kajiwara, M. Noumi, Y. Yamada, *Geometric aspects of Painlevé equations*, J. Phys. A: Math. Theor. **50** (2017), no. 7, 073001 (164 pp.).

[97] K. Kajiwara, Y. Ohta, *Determinant structure of the rational solutions for the Painlevé II equation*, J. Math. Phys. **37** (1996), 4393–4704.

[98] K. Kajiwara, Y. Ohta, *Determinant structure of the rational solutions for the Painlevé IV equation*, J. Phys. A: Math. Gen. **31** (1998), no. 10, 2431–2446.

[99] A.A. Kapaev, *Quasi-linear Stokes phenomenon for the Painlevé first equation*, J. Phys. A: Math. Gen. **37** (2004), 11149–11167.

[100] A.V. Kitaev, C.K. Law, J.B. McLeod, *Rational solutions of the fifth Painlevé equation*, Diff. Integral Equations **7** (1994), 967–1000.

[101] R. Koekoek, P.A. Lesky, R.F. Swarttouw, *Hypergeometric orthogonal polynomials and their q-analogues*, Springer Monographs in Mathematics, Springer-Verlag, Berlin, 2010.

[102] A.B.J. Kuijlaars, K.T.-R. McLaughlin, *Generic behavior of the density of states in random matrix theory and equilibrium problems in the presence of real analytic external fields*, Comm. Pure Appl. Math. **53** (2000), no. 6, 736–785.

[103] A.B.J. Kuijlaars, K.T.-R. McLaughlin, *Asymptotic zero behavior of Laguerre polynomials with negative parameter*, Constr. Approx. **20** (2004), 497–523.

[104] A.B.J. Kuijlaars, R. Milson, *Zeros of exceptional Hermite polynomials*, J. Approx. Theory **200** (2015), 28–39.

[105] A.B.J. Kuijlaars, W. Van Assche, *Extremal polynomials on discrete sets*, Proc. London Math. Soc. (3) **79** (1999), 191–221.

[106] E. Laguerre, *Sur la réduction an fractions continues d'une fraction qui satisfait à une équation différentielle linéaire du premier ordre dont les coefficients sont rationnels*, J. Math. Pures Appl. (4) **1** (1885), 135–166.

[107] J.S. Lew, D.A. Quarles, Jr., *Nonnegative solutions of a nonlinear recurrence*, J. Approx. Theory **38** (1983), 357–379.

[108] D.S. Lubinsky, H.N. Mhaskar, E.B. Saff, *A proof of Freud's conjecture for exponential weights*, Constr. Approx. **4** (1988), no. 1, 65–83.

[109] N.A. Lukashevich, *On the theory of the third Painlevé equation*, Diff. Uravn. **3** (1967), no. 11, 1913–1923 (in Russian); translated in Differ. Equations **3** (1967), 994–999.

[110] S. Lyu, Y. Chen, *Exceptional solutions to the Painlevé VI equation associated with the generalized Jacobi weight*, Random Matrices Theory Appl. **6** (2017), no. 1, 1750003, 31 pp.

[111] A.P. Magnus, *A proof of Freud's conjecture about the orthogonal polynomials related to $|x|^\rho \exp(-x^{2m})$, for integer m*, in "Orthogonal Polynomials and Applications" (Bar-le-Duc, 1984), Lecture Notes in Mathematics **1171**, Springer, Berlin, 1985, pp. 362–372.

[112] A.P. Magnus, *Freud's equations for orthogonal polynomials as discrete Painlevé equations*, in "Symmetries and Integrability of Difference Equations", Canterbury 1996, London Math. Soc. Lecture Note Series **255**, Cambridge University Press, 1999, pp. 228–243.

[113] A.P. Magnus, *Painlevé-type differential equations for the recurrence coefficients of semi-classical orthogonal polynomials*, J. Comput. Appl. Math. **57** (1995), 215–237.

[114] A.P. Magnus, *Special nonuniform lattice (snul) orthogonal polynomials on discrete dense sets of points*, J. Comput. Appl. Math. **65** (1995), no. 1–3, 253–265.

[115] A.P. Magnus, *Freud equations for Legendre polynomials on a circular arc and solution of the Grünbaum-Delsarte-Janssen-Vries problem*, J. Approx. Theory **139** (2006), 75–90.

[116] D. Masoero, P. Roffelsen, *Poles of Painlevé IV rationals and their distribution*, arXiv:1707.05222 [math.CA] (July 2017).

[117] T. Masuda, *Classical transcendental solutions of the Painlevé equations and their degeneration*, Tohoku Math. J. **56** (2004), 467–490.

[118] T. Masuda, Y. Ohta, K. Kajiwara, *A determinant formula for a class of rational solutions of Painlevé V equation*, Nagoya Math. J. **168** (2002), 1–25.

[119] A. Máté, P. Nevai, T. Zaslavsky, *Asymptotic expansions of ratios of coefficients of orthogonal polynomials with exponential weights*, Trans. Amer. Math. Soc. **287** (1985), no. 2, 495–505.

[120] M. Mazzocco, *Rational solutions of the Painlevé VI equation*, J. Phys. A: Math. Gen. **34** (2001), no. 11, 2281–2294.

[121] P.D. Miller, Y. Sheng, *Rational solutions of the Painlevé-II equation revisited*, Symmetry Integr. Geom. Meth. Appl. (SIGMA) **13** (2017), 065, 29 pp.

[122] P. Nevai, *Orthogonal polynomials associated with $\exp(-x^4)$*, in "Second Edmonton Conference on Approximation Theory", Canadian Math. Soc. Conf. Proc. **3** (1983), pp. 263–285.

[123] NIST Digital Library of Mathematical Functions, http://dlmf.nist.gov/, Release 1.0.11 of 2016-06-08. Online companion to [137].

[124] A.F. Nikiforov, S.K. Suslov, V.B. Uvarov, *Classical orthogonal polynomials of a discrete variable*, Springer Series in Computational Physics, Springer-Verlag, Berlin, 1991.

[125] M. Noumi, *Painlevé equations through symmetry*, Translations of Mathematical Monographs, vol. 2233, Amer. Math. Soc., 2004.

[126] M. Noumi, S. Okada, K. Okamoto, H. Umemura, *Special polynomials associated with the Painlevé equations, II* in "Integrable Systems and Algebraic Geometry" (Kobe/Kyoto, 1997), World Scientific, River Edge, NJ, 1998, pp. 349–372.

[127] M. Noumi, Y. Yamada, *Umemura polynomials for the Painlevé V equation*, Phys. Lett. **A247** (1998), 65–69.

[128] M. Noumi, Y. Yamada, *Symmetries in the fourth Painlevé equation and Okamoto polynomials*, Nagoya Math. J. **153** (1999), 53–86.

[129] V.Yu Novokshenov, A.A. Shchelkonogov, *Double scaling limit in the Painlevé IV equation and asymptotics of the Okamoto polynomials*, in "Spectral theory and differential equations" (V.A. Marchenko's 90th anniversary collection), Amer. Math. Soc. Transl. Ser. 2, **233**, Providence, RI, 2014, pp. 199–210.

[130] V.Yu. Novokshenov, A.A. Schelkonogov, *Distribution of zeros of generalized Hermite polynomials*, Ufimskii Mat. Zhurnal **7** (2015), no. 3, 57–69 (in Russian); translated in Ufa Math. J. **7** (2015), no. 3, 54–66.

[131] Y. Ohyama, H. Kawamuko, H. Sakai, K. Okamoto, *Studies on the Painlevé equations, V: Third Painlevé equations of special type* $P_{III}(D_7)$ *and* $P_{III}(D_8)$, J. Math. Sci. Univ. Tokyo **13** (2006), 145–204.

[132] K. Okamoto, *Sur les feuilletages associés aux équations du second ordre à points critiques fixés de P. Painlevé*, Japan J. Math. (N.S.) **5** (1979), 1–79.

[133] K. Okamoto, *Studies on the Painlevé equations I. Sixth Painlevé equation* P_{VI}, Ann. Mat. Pura Appl. **146** (1987), 337–381.

[134] K. Okamoto, *Studies on the Painlevé equations II. Fifth Painlevé equation* P_V, Japan J. Math. **13** (1987), 47–76.

[135] K. Okamoto, *Studies on the Painlevé equations III. Second and fourth Painlevé equations,* P_{II} *and* P_{IV}, Math. Ann. **275** (1986), 221–255.

[136] K. Okamoto, *Studies on the Painlevé equations IV. Third Painlevé equation* P_{III}, Funkcial. Ekvac. **30** (1987), 305–332.

[137] F.W.J. Olver, D.W. Lozier, R.F. Boisvert, C.W. Clark (eds.), *NIST handbook of mathematical functions*, Cambridge University Press, New York, NY, 2010. Print companion to [123].

[138] V. Periwal, D. Shevitz, *Unitary-matrix models as exactly solvable string theories*, Phys. Rev. Letters **64** (1990), 1326–1329.

[139] I.E. Pritsker, R.S. Varga, *The Szegő curve, zero distribution and weighted approximation*, Trans. Amer. Math. Soc. **349** (1997), no. 10, 4085–4105.

[140] R. Sasaki, S. Tsujimoto, A. Zhedanov, *Exceptional Laguerre and Jacobi polynomials and the corresponding potentials through Darboux-Crum transformations*, J. Phys. A: Math. Theor. **43** (2010), no. 31, 315204, 20 pp.

[141] H. Segur, M. Ablowitz, *Asymptotic solutions of nonlinear evolution equations and a Painlevé transcendent*, Phys. D **3** (1981), no. 1–2, 165–184.

[142] J. Shohat, *A differential equation for orthogonal polynomials*, Duke Math. J. **5** (1939), no. 2, 401–417.

[143] B. Simon, Orthogonal polynomials on the unit circle, Amer. Math. Soc. Colloq. Publ. **54**, Part 1 and Part 2, Amer. Math. Soc., Providence, RI, 2005.

[144] C. Smet, W. Van Assche, *Orthogonal polynomials on a bi-lattice*, Constr. Approx. **36** (2012), no. 2, 215–242.

[145] G. Szegő, *Über eine Eigenschaft der Exponentialreihe*, Sitzungsber. Berl. Math. Ges. **23** (1924), 50–64.

[146] G. Szegő, *Orthogonal polynomials*, Amer. Math. Soc. Colloq. Publ., vol. 23, Amer. Math. Soc., Providence, RI, 1939 (4th edition 1975).

[147] C.A. Tracy, H. Widom, *Random unitary matrices, permutations and Painlevé*, Commun. Math. Phys. **207** (1999), no. 3, 665–685.

[148] H. Umemura, *Special polynomials associated with the Painlevé equations, I*, manuscript (presented at the workshop on the Painlevé transcendents, Montréal, 1996).

[149] W. Van Assche, *Discrete Painlevé equations for recurrence coefficients of orthogonal polynomials*, in "Difference Equations, Special Functions and Orthogonal Polynomials" (S. Elaydi, eds.), World Scientific, 2007, pp. 687–725.

[150] W. Van Assche, S.B. Yakubovich, *Multiple orthogonal polynomials associated with Macdonald functions*, Integral Transform. Spec. Funct. **9** (2000), no. 3, 229–244.

[151] A.P. Vorobiev, *On rational solutions of the second Painlevé equation*, Differ. Uravn. **1** (1965), 79–81 (in Russian); translated in Differ. Equations **1** (1965), 58–59.

[152] N.S. Witte: *Semiclassical orthogonal polynomial systems on nonuniform lattices, deformations of the Askey table, and analogues of isomonodromy*, Nagoya Math. J. **219** (2015), 127–234.

[153] S.-X. Xu, D. Dai, Y.-Q. Zhao, *Critical edge behavior and the Bessel to Airy transition in the singularly perturbed Laguerre unitary ensemble*, Comm. Math. Phys. **332** (2014), 1257–1296.

[154] S.-X. Xu, D. Dai, Y.-Q. Zhao, *Painlevé III asymptotics of Hankel determinants for a singularly perturbed Laguerre weight*, J. Approx. Theory **192** (2015), 1–18.

[155] Shuai-Xia Xu, Yu-Qiu Zhao, *Painlevé XXXIV asymptotics of orthogonal polynomials for the Gaussian weight with a jump at the edge*, Stud. Appl. Math. **127** (2011), no. 1, 67–105.

[156] A.I. Yablonskii, *On rational solutions of the second Painlevé equation* (in Russian), Vestsi Akad. Nauvuk BSSR, Ser. Fiz. Tekh. Navuk **3** (1959), 30–35.

Index